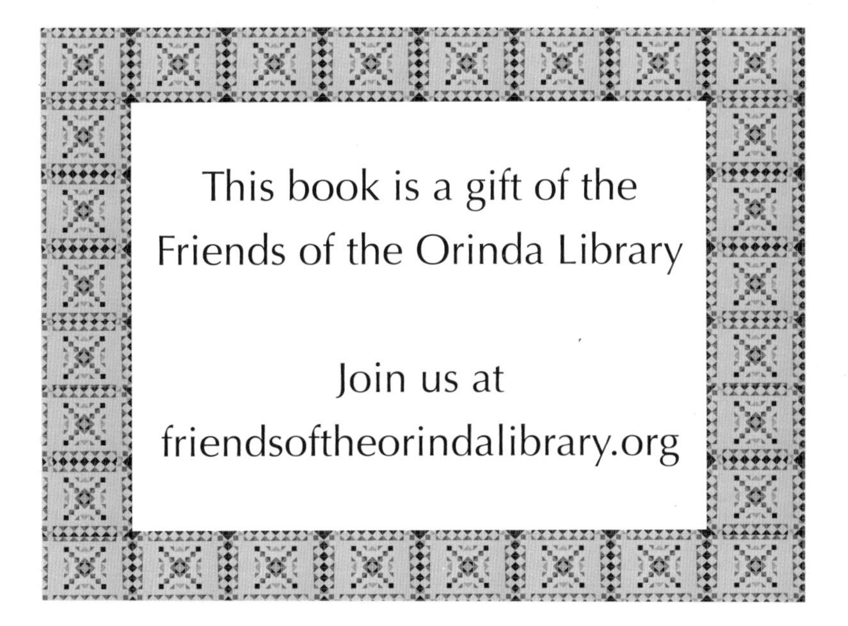
This book is a gift of the
Friends of the Orinda Library

Join us at
friendsoftheorindalibrary.org

COWED

Also by Denis Hayes

Rays of Hope: The Transition to a Post-Petroleum World
The Official Earth Day Guide to Planet Repair
Pollution: The Neglected Dimensions
Repairs, Reuse, Recycling: First Steps Toward a Sustainable Society

Also by Gail Boyer Hayes

Pulmonary Hypertension: A Patient's Survival Guide
Solar Access Law

COWED

The Hidden Impact of
93 Million Cows on America's Health,
Economy, Politics, Culture, and Environment

Denis Hayes &
Gail Boyer Hayes

W. W. NORTON & COMPANY
New York · London

Printed in the United States of America
First Edition

For information about special discounts for bulk purchases, please contact W. W. Norton Special Sales at specialsales@wwnorton.com or 800-233-4830

Manufacturing by RR Donnelley, Harrisonburg
Book design by Kristen Bearse
Production manager: Anna Oler

ISBN: 978-0-393-23994-2

W. W. Norton & Company, Inc., 500 Fifth Avenue, New York, N.Y. 10110
www.wwnorton.com

W. W. Norton & Company Ltd., Castle House, 75/76 Wells Street, London W1T 3QT

1 2 3 4 5 6 7 8 9 0

For our daughter, Lisa, our granddaughter, Sheridan,
. . . and for cows.

Contents

Introduction

SETTING OUT

Cows shaped America. Cows enabled Europeans to successfully occupy the Continent. Cows, and the food grown to feed them, radically remodeled the nation's landscape. Cows are partly responsible for Americans' increasingly rotund bodies and poor health. Cows have exerted a remarkable degree of influence over our economic system, our politics, and our culture. Although cows themselves are now mostly hidden out of sight, cow molecules abound in every room of our houses. If gathered up and put on one side of a giant balancing scale, with humans on the other side, the cows living in the United States would weigh two and a half times as much as the human population.

Cows matter.

Our focus on cows began as a lark some years ago when we drove through Scotland, Ireland, Wales, and England. Leaving Edinburgh, we were immediately struck by the many small herds we saw. In the United States, even in farm country, we seldom see cows these days—they have been moved off pastures and into confined feeding lots or giant dairy barns far from thoroughfares. Furthermore, the cows we saw in the British Isles appeared to come in more varieties. To add spice to the trip, we began competing to spot, and snap photos of, unusual cows.

Near the end of our trip, we had reservations at a thatched-roofed

B&B deep in the verdant countryside of Devon, England. The tangle of one-lane country roads and the paucity of road signs prevented us from meeting our hostess till dusk. But there was still time for a short walk before dinner. Gail—who reliably gravitates toward any warm-blooded animal—had noticed a small herd of Holsteins grazing in a nearby pasture. White hides sporting Rorschach-like inkblots gave the cows a festive air.

We strolled down the hill toward them, picking our way around stones, bushes, and cow pies. Fifty feet from the cows, we paused, a flimsy hedge separating us from them.

The nearest cow was snacking on a tree. A tongue rolled out of its mouth like a muscular tsunami, wrapped itself around a small branch, and reeled the branch in. We edged closer. When we were about thirty feet away, the cow stopped eating and stared at us, leaves sticking out of both sides of its mouth. This cow was huge, easily five times as large as both of us combined. As city dwellers, we tend to forget just how large cows are. Relatively few Americans ever have a face-to-face encounter with the source of their milk and hamburgers. A trivial fraction of 1 percent of Americans has had the experience of milking a cow; an even smaller fraction has killed a cow to eat its flesh. We are more isolated from the source of our food than any previous generation.

The Holstein's tail switched back and forth. It wasn't a friendly sort of puppy-tail wagging or a lazy shoofly motion. The half-dozen other cows also stopped snacking and all of them stared at us. Not knowing any better, we stared back. In the air between us hung a bubble of mutual awareness.

Unlike horses, cows don't raise their heads high. The lowered heads, combined with unblinking stares, seemed menacing. Cows in those parts had no reason to trust or love humans. Over *four million* of them had recently been slaughtered and burned to prevent the spread of mad cow disease.

We knew that Holsteins were well regarded as milk cows, each cow now producing twice as much milk as a cow did in the 1960s, when we were in college. Our decision to stroll into the pasture had been prompted, in part, by our idle curiosity about these prodigious modern milk machines. The nearest cow pawed the ground. Gail nudged Denis and whispered, "No udders. They must be bulls." Unlike the hornless Holsteins pictured on milk cartons, these animals had sturdy, sharp-tipped horns.

Denis squinted at the herd. "Nope. They're steers." Still, a steer is no pushover.

From where we stood we couldn't tell if there was a fence buried in the hedgerow between us and the steers, or whether the barrier consisted only of delicate twigs and leaves. The lit windows of the B&B high behind us seemed far away in the gathering darkness. We backed off and returned to our lodgings.

Gail's interest in cows outlasted the trip. Back home in Seattle, she began reading up on cows. Her conversation became peppered with cow facts. Did Denis know, for example, that cows sweat through their noses? That they have a magnetic sense and tend to align themselves in a north/south direction when grazing?[1] On drives, John Pukite's *A Field Guide to Cows* in hand, she started watching and identifying breeds of cows the way some people watch birds.

It was hard to find many cows to watch, however. Were the hidden-away cows being treated well? Books and articles made it clear that many factory-farmed cows were abused. She wondered whether it was possible, in the States, to catch a brain-eating disease from eating a well-cooked hamburger. What she learned was not reassuring.

Denis, with his broad interest in matters environmental and a penchant for hiking around obscure corners of the world, had always been more interested in endangered wild animals than in domesticated creatures. Eventually, though, he came to share Gail's convic-

From a tourist's perspective, there appears to be a greater variety of cows in Britain than one sees from roads in the United States. Highland cow, Scotland. Photo: Lyda Boyer.

tion that cows deserve more attention. Once he focused on cows, he found their heavy hoofprints on almost every major environmental problem. Compared to other meat sources (like pork and poultry), conventional grain-finished feedlot beef produces five times more global warming per calorie, requires eleven times more water, and uses twenty-eight times as much land. Eating a pound of beef has a greater climatic impact than burning a gallon of gasoline.[2] And as he looked more carefully at feedlots and conventional dairy practices, Denis, too, became ashamed of how Americans treated cows.

We both sensed that if we could just connect the dots, an important story might be revealed about the impact of America's ninety-three million cows[3]—roughly one hundred and twenty *billion* pounds of cow—on our lives. We began to spend evenings reading about the curiously entangled lives of cows and Americans and discovered that cows cast considerable light on who we are as a people.

Problems caused by cows were easy to find. Solutions were more elusive. We left Seattle to seek out people who treat their cows well, who are finding solutions to cow-related environmental problems, and who are applying lessons learned from eons of beta-testing by Mother Nature. Among them were beef ranchers and worm wranglers, dairy farmers, an artisanal cheese maker, soil experts, and public health professionals.

People and cows have been modifying each other's genomes and environments for thousands of years. Initially, those changes benefited both species. More recently, the costs to humans and cows alike outweigh the benefits. Aided and abetted by a shortsighted federal farm policy, Big Ag (corporate agriculture) treats cows barbarously, even as it ruins some of the best soil on the planet, destroys irreplaceable aquifers, fills the air with warming gasses, and creates enormous dead zones at the mouths of rivers.

All this was interesting. But the two of us are getting on in years, and still hoping to clean up some of the mess we'll otherwise leave behind for future generations. Was writing about cows a good way to spend our time? We decided it would be . . . *if* we could propose some solutions. Convincing readers that something needs to be done should be easy, we reasoned. Getting people to act would be much harder. We looked around for trends already under way. It helps to have the wind at your back. We found those trends in the popularity of the organic and eat-local movements, in the mounting concern over food security in an era of climate disruption, and in the increasing urgency to do something about the unhealthy diets of many Americans.

Our proposal is affordable and simple. It does not require the engagement of America's dysfunctional Congress or its compromised bureaucracies. It does not require civil disobedience, mass marches, or expensive lawsuits. All it requires is that enough like-minded people seek out organic dairy products and grass-fed-and-finished beef. Most people can do this without busting their budgets by reducing their beef and dairy consumption to levels that are better for their health. As with computers, smartphones, and wheeled suitcases, so will it be with ranches and dairies: When upstarts begin eating the economic lunch of the establishment, even the most reluctant establishment must change—or become history's roadkill.

Chapter One

BEWITCHED BY COWS

Some of the earliest paintings, made over thirty thousand years ago, are in Chauvet Cave in France. They depict a remarkable creature that had shoulders hefty as a Brink's truck and long, lyre-shaped horns with a purposeful forward slant. Called aurochs, these animals prance over cave walls on delicate-looking legs. The paintings prove that humans' fascination with bovines began long before we domesticated them.

Aurochs (pronounced OR-ox) were formidable. Julius Caesar wrote, "Their strength and speed are extraordinary; they spare neither man nor wild beast which they have espied." Roman gladiators fought aurochs in the Colosseum. "But not even when taken very young can they be rendered familiar to men and tamed," Caesar wrote.[1]

Caesar was wrong. This majestic beast was the ancestor of the modern cow and its domestication had already begun thousands of years earlier in Persia and, separately, in India. Aurochs could outrun, outfight, and generally outcompete today's cows in any competition but the production of milk and marbled meat. ("Marbling" is a Madison Avenue word for "fat.") On marshy land they could even outrun horses, because they had cloven hooves. (Wider hooves don't sink as fast, and a space between toes lets air out so vacuums don't form.)

At one time aurochs roamed at will over most of Europe and Central Asia, as well as parts of North Africa, the Middle East, and India. That's more territory than even the most assertive human has ever managed to conquer. Male aurochs were black with a pale stripe down their spine; females and calves were a fetching reddish hue. A bull could be six feet tall at his shoulders and weigh over a ton.

Aurochs were driven to extinction by loss of habitat to farming and by domesticated cattle that competed with them for food and infected them with diseases. *Homo sapiens* inflicted the coup de grâce. In addition to eating aurochs, having a fine collection of silver-tipped aurochs-horn cups to drink from implied nice things about one's status and manhood. Aurochs held out longest in the forests of Poland because only Polish royalty were allowed to hunt them. When their numbers had dwindled sufficiently, even the king declined to kill them. But by then it was too late. The last aurochs died in Poland in 1627. Her skull is now in a museum in Stockholm.

Cows and Humans Remodel Each Other

Foul-tempered, oversized beasts weren't of much use to farmers. So about ten thousand years ago, as humans became increasingly agrarian, they began to pick out and breed those aurochs that were smaller and more tractable—a process of unnatural selection. It wasn't easy to domesticate aurochs, and genetic sleuthing suggests it was rarely successful: All the "taurine" cattle, the cows we see around us in the United States, can be traced back to about eighty biddable female aurochs that were domesticated by ancient Persians. Humped cattle were domesticated separately, from a subspecies of aurochs on the Indian subcontinent.

A cow with a laid-back personality is not only easier to handle but gains weight faster. It didn't hurt that the aurochs selected for their gentle nature were also a bit dumber than other herd members. Genes heavily influence temperament. Scientists recently learned that regions of the cow genome that were unknowingly altered by those early breeders include genes that—in the human genome—are linked to autism, mental retardation, and general brain development.[2]

Over time, humans further selected cows with all sorts of special traits, such as heavy milk production, tasty meat, winsome looks, the strength to pull heavy loads, and the ability to survive in various environments. Today there are more than eight hundred recognized breeds[3] plus a great many mongrels. Cows can become "friends with benefits" with different cow breeds and even with close relatives such as bison, yak, banteng, and gaur. Nevertheless, many potentially valuable breeds are now headed toward extinction.[4]

Of course, from cows' point of view, they domesticated humans. Those cows that found humans who would protect them from predators and provide them with food and water survived and multiplied, eventually outnumbering aurochs. Humans later transported cows to distant continents where cows thrived and their numbers swelled. Today, humans devote an extraordinary percentage of available land, water, grain, and energy to meeting the needs of cows.

While humans were busy changing the cows' genome, cows simultaneously changed the human genome. In the Stone Age, the only humans able to tolerate milk were babies, because babies make lactase, an enzyme needed to metabolize sugars in their mother's milk. Ötzi, the 5,300-year-old "Tyrolean Iceman" whose frozen body was uncovered in the Italian Alps in 1991, was lactose intolerant; cows were still in the process of changing Europeans' genome when Ötzi lived. Humans who were better able to utilize available nutrients, including milk, were more likely to survive famines and perhaps had more energy to go out and conquer other cultures.

In their provocative book *The 10,000 Year Explosion*, Gregory Cochran and Henry Harpending argue that it may also be thanks to cows that half of mankind now speaks an Indo-European language. Their reasoning goes this way: Humans who could tolerate lactose (in particular the Kurgan people of Southwest Asia) were more mobile than farmers, because cows are easier to move around than crops. They were also better-fed, because it's possible to get five times the calories per acre from cows when they are used for milk rather than meat. More warriors therefore could be raised on the same amount of land, and those warriors could hone their skills by stealing cows from other mobile tribes. When attacked by farmers, the milk drinkers could just retreat (along with their cows). When milk drinkers attacked farmers, the farmers couldn't leave their crops, had fewer warriors, and often lost women to the raiders (along with opportunities to procreate). Thus the genetic alteration that allowed humans to consume milk spread, along with the language of the milk drinkers.[5]

And now the story of cows comes full circle. Around 1920, the Heck brothers of Germany began to breed domesticated cattle to resemble their mighty aurochs ancestor. The Nazi government supported this effort because it fit in well with propaganda about a magnificent Aryan race. Today there are about two thousand Heck cattle that somewhat resemble their fierce forbears, although they aren't as big. A Dutch foundation, Stichting Taurus, has joined forces with the Ark Foundation and Rewilding Europe to try to produce aurochs by selectively breeding the existing cattle that most resemble aurochs. The goal of this ambitious effort is to create ten large wild areas in Europe. The ecology of such areas requires the presence of large grazing animals capable of defending themselves against wolves and other predators. Aurochs filled that ecological niche for over 250,000 years.

Poland, home of the last aurochs, is trying to bring back the beast more directly. Geneticists recently dug aurochs DNA out of museum samples with the goal of creating living, breathing aurochs. A cross-disciplinary group of Polish scientists is hard at work on this project.[6] If they succeed, and create an Aurochs Park, you'll need a tank to drive through it.

Cows Shape America's Landscape

In his thoughtful book *Guns, Germs, and Steel*, Jared Diamond attributes the ease of European colonial conquest of America to Europeans' possession of guns and steel, and to the diseases Europeans carried in their bodies, diseases to which indigenous populations had not developed immunity. To this list we would add another key factor: Europeans had cows.[7]

Two broad categories of cows were brought to the New World: gentle, carefully bred cows from England and places close to it, and Iberian cattle flashing long horns and lean loins. English cows landed in North America on the East Coast; Spanish cows landed farther south and came up through Mexico.

In 1607 a ship named the *Susan Constant* brought cows to Jamestown, Virginia. Unfortunately, these cows were all eaten in the terrible winter of 1610, but more arrived in 1611. Although no cows can claim a *Mayflower* pedigree, the Plymouth Colony settlers, who landed late in 1620, received a shipment of cows by 1623. A decade later, the Massachusetts herd had grown to 1,500 cows. These East Coast cows had been bred to be productive and malleable. They were carefully tended, sheltered, and fed hay by their owners when fresh grass was scarce.

In Texas, Arizona, New Mexico, and along the California coast, Franciscan and Jesuit priests operated missions from the 1600s through the early 1800s. Each mission had satellite ranches well stocked with Iberian longhorns. For example, the San Luis Rey mission had an impressive four ranches in 1832 and 27,500 cattle. The priests lured Native American converts by offering them beef, and then trained them to help manage the cattle.[8] The tough longhorns weren't coddled. Used mostly for meat, tallow, and hide, they took care of themselves in unfenced pastures. Periodically the longhorns would be rounded up and branded or slaughtered. Bulls often weren't castrated, and they were formidable foes for predators. In subtropical areas of the Texas and Louisiana coasts, longhorns fed on lush grasses without supplemental food.[9] Longhorns didn't have barns.

Eventually, feral Iberian cows (cows that had such minimal contact with humans that they were almost wild) in Texas bred with gentler cows from the East. Their progeny, the Texas longhorns, were meatier and more tractable than Iberians yet still had good survival skills. The rapid expansion of Texas longhorns and Texas-style ranching onto the Great Plains after the Civil War coincided with an unusually warm and wet spell. Ranchers overestimated how many cattle the plains could support. Twenty million cattle, in herds of about three thousand head, each with a crew of about ten cowboys, were driven from Texas to railheads in Missouri, Kansas, Nebraska, and Wyoming. This was the time of the great cattle drives along the Shawnee, Chisholm, Western, and Goodnight-Loving trails—drives that inform folklore and fill movie theaters.

But the weather reverted to normal, the grass dried up, and the resulting overgrazing compacted soils and destroyed perennial grasses that cows favored. The loss of vegetation led to more evaporation, erosion, and even desertification. The summer of 1886 saw temperatures as high as 110 degrees Fahrenheit; the following winter temperatures plunged to –45 degrees. Unable to tolerate such extremes, half the cattle on the Great Plains died.

By 1890, Texas-style open-range cattle ranching had collapsed. A new breed of settler from the East was increasing in numbers and influence. These settlers had barns, bales of hay, and barbed wire fencing (patented in 1874). The cow-punching drifters of legend disappeared along with the longhorns. "In the long run, only the myth survived," writes historian Terry G. Jordan.[10]

Cow owners who used eastern methods on the plains suffered smaller losses when the wet weather ended because they invested far more time and money in taking care of their cows. Their equestrian and lasso-tossing skills weren't impressive, and they didn't fit the romantic image of the West that Hollywood uses to attract moviegoers. But they were successful farmers, and they created the template for today's massive cattle industry.

Settlers who moved to America from England carried in their minds an image of the ideal farm, a place suited to the damp, cool English climate. As they moved westward, they replaced many native grasses and other plants with imported crops to feed themselves and their cows. The English approach worked fairly well in many parts of the eastern United States. Thomas Jefferson became famously rhapsodic on the virtues of the family farmer (that is, people like himself): "Those who labor in the earth are the chosen people of God . . . whose breasts He has made His peculiar deposit for substantial and genuine virtue."[11]

Having survived a dark history of feudalism, most colonists and their offspring extolled the virtues of individual ownership of land. True, the weather in what would become the United States was more violent than Old Country weather; thunderstorms and temperature fluctuations were hard on soil. But when settlers mined the heart out of the soil in one place, they just moved west.

After the Civil War, vast new opportunities opened up for settlers to acquire land. The federal government was located hard on the East Coast, and for a long time policymakers had little sense of what the lands to the west of eastern Ohio were actually like. Geologist explorer John Wesley Powell led expeditions to investigate, and in 1878 wrote his prescient *Report on the Lands of the Arid Region*. Congress blithely ignored Powell's advice to designate some lands as suitable only for pasture or very large ranches and instead divided the arid region into 160-acre quarter sections.

In 1912 Congress expanded this to as much as 640 acres, which is one square mile. To urban apartment dwellers, that may seem generous, but ranchers in arid lands need far more acreage than that to earn a living. Powell had advised that ranches be *at least* 2,560 acres.[12]

With the buffalo essentially exterminated, the native people subdued, and the new opportunity to own 160 acres of land by homesteading, the great land rush began. The rush attracted even more European immigrants and their cows. In 1885, Norman Coleman, the first United States Commissioner of Agriculture, bragged: "If a solid column should be formed, twelve animals deep, one end resting at New York City, its centre encircling San Francisco, and its other arm reaching back to Boston, such a column would contain about the number which now forms the basis, the capital stock, so to speak, of the cattle industry of the United States." Try to visualize that. Then consider that the United States had 45 million cows in 1885.[13] We have double that number today.

• • •

"In the most stupendous migration of recorded history the great inseparables were the immigrant and the ox," wrote historians Charles Wayland Towne and Edward Norris Wentworth.[14] An ox isn't a separate species. It's just a castrated bull, or occasionally a female cow, that has been selected and trained to work as a draft animal. Many breeds can make fine oxen. In America, teamsters (men who work oxen) usually pick Chianina, Dutch Belted, Holstein, Jersey, Brown Swiss, Milking Devon, or Milking Shorthorn.[15] Oxen usually work in pairs, joined by a yoke. Pioneers preferred oxen to pull their covered wagons because oxen were stronger and steadier than horses, and a yoke allowed them to pull far heavier loads. Oxens' strong, steady pace also made them superior plow animals for clay or rocky soils. Oxen enabled one man to plow enough acreage to grow more than enough food to feed himself and his family. He could sell his surplus crops or trade them for goods. As a bonus, oxen fertilized his fields. And they were powered by grass. This was a win/win/win setup.

Freight caravans employed vast numbers of oxen. The Russell, Majors and Waddell firm—the United Van Lines of the oxen era—owned seventy-five thousand oxen.[16] Oxen also greatly facilitated the spread of logging across the northern tier of the nation. Babe the Blue Ox, the companion of folklore logger Paul Bunyan, symbolized this role. Even today, ox logging is an important, if boutique, part of some sustainable forestry operations. There are still many oxen in the New England region of America, found mostly on smaller farms and maintained as a hobby.[17]

Ultimately, of course, machines replaced nearly all oxen in farming, logging, and transport. Railroads provided a key link in the growing beef industry, hauling cattle from connection points like

Abilene to meatpacking centers like Chicago. Instead of providing the means of transportation, cows became commodities to be shipped.

Cows Shape American Culture

Cows shaped our culture as thoroughly as they did our landscape. Literature (westerns), movies (westerns), dance (line dancing, square dancing), heroes (Wyatt Earp, Davy Crockett, Wild Bill Hickok) and villains (Billy the Kid, Jesse James), architecture (the ranch house), diet (steaks and hamburgers), games (cowboys and Indians), vehicles (SUVs and pickups), addictive habits (Marlboro cigarettes and Red Man chewing tobacco)—all these were heavily influenced by cowboy mythology. Cows and cowboys have had a particularly heavy impact on television. We counted more than 170 western series, some running a decade or longer. These include *Bonanza, Daniel Boone, Dallas, Davy Crockett, Death Valley Days, Deadwood, Have Gun—Will Travel, The Life and Legend of Wyatt Earp, The Lone Ranger,* and *Wagon Train*. All these were heavily influenced by cowboy mythology. It's such a seductive ideal that it has spread beyond our boundaries and naturalized in the cultures of many other nations.

Beef cattle have done more to define what it means to be American than have dairy cows (sorry, Jefferson). And it was beef cattle, with their horns and a wicked glint in their eye, that elevated cowboys to role model status. Real cowboys came from the lower stratum of society and labored hard and long, in nasty weather, for little pay. So how did a caste of hired (and sometimes enslaved) laborers, who often didn't even own their own horses, become cultural heroes?

The cowboy myth really caught on after the Civil War. A nation deeply divided was in desperate need of heroes both sides could agree on, and cowboys fit the niche. Here is researcher Jennifer Moskowitz's take:

> [T]he country needed a unifying, nationalist icon to move it beyond the ravages of the Civil War and the Englishness of Southern agrarian society into industrialism and capitalism. . . . Into the West rode the American cowboy, whose mythic figure and setting were equally significant and carefully shaped by authors, artists, and political figures. [I]ronically . . . the qualities ascribed to the cowboy are identical to those of the English knight.[18]

The cowboy of modern mythology is not wholly devoid of grounding in reality. But Hollywood and Marlboro felt a need to embellish the unshowered, illiterate men with short tempers and bad teeth. Although movie cowboys speak English, many of the real cowboys came from the Spanish/Mexican vaqueros. Others were Native Americans, such as Cherokees dispossessed of their land, and former Negro slaves released by the Civil War. The term "cowpoke" was used before the word "cowboy," because Hispanic cattle herders poked cows with blunt lances.[19] Neither "cowboy" nor "cowpoke" has much inherent romance as a word, but a person who poked cows with a stick for a living might have been beyond even the power of Hollywood to romanticize.

Let's disassemble the elements of the cowboy. Something can be so obvious that it is rendered out-of-mind and almost invisible: These knights with shiny belt buckles were almost exclusively male. It took physically strong men to tackle the challenges of the frontier and control big animals. Texans used the derogatory term "she-rancher" to describe a man who made the more profitable decision to stay home and care for his cattle during the winter.[20]

Being a cowboy meant being willing to take risks. Studies have found an association between high testosterone levels and risky behavior.[21] Testosterone levels are highest in young men. Historian David T. Courtwright notes that until World War II there were more men than women in the United States. A hefty percentage of men were young, especially those on the frontier. Courtwright theorizes that this imbalance between the sexes set in motion a culture of testy violence and aggression,[22] echoes of which still resound today.

Having a horse also mattered. Knight or cowboy, a horse lifts a man into the mythic realm. When mounted, men appear less dwarfed by a big-sky landscape and potentially more dangerous. (This is probably why so many statues and paintings insert a horse under the hero.) Unlike cows, horses hold their heads high, looking proud. It takes skill to ride a horse well. People on foot must literally look up to someone riding a horse.

Like cows, horses are very large animals, and size matters to cowboys. While researching this book, we talked with a Montana rancher who had experimented with raising Australian Lowline cattle instead of standard Angus cows. His cowboys rebelled—they just didn't feel right bossing around sweet-natured, hornless, three-and-a-half-foot-tall cows. Lowlines are wonderful mothers, ideal for intensive grazing, and easy to raise. They produce a higher percentage of usable meat than taller cows, and their meat is nicely marbled with a minimum of grain in their diet. But herding them makes cowboys feel ridiculous.

The cowboy ethos has persisted because, along with what many consider antiquated elements, there is much to admire. A key element of the myth is that hard work and initiative will pay off. And for pioneers it often did. The frontier gave cowboys and settlers alike a degree of freedom from being controlled by other people. Out on the range, no midlevel manager was at hand to micromanage how the cowboy wound his lasso or to monitor his bathroom breaks. The cowboy lived hard against the elements, and stupidity was quickly and impersonally punished. He was free to use the land as he wished. There was no time to form a committee to reach a consensus on whether to shoot a wolf or put down a cow with a broken leg. A cowboy had to make snap decisions. He lived in a world of white hats and black hats, no shades of gray.

The nineteenth-century cowboy, out on the trail for months at a time with just his gun and his horse, was also a model of self-reliance. He was the sort of person Frederick Jackson Turner had in mind when he developed his famous thesis, "The Significance of the Frontier in American History," which he presented in 1893 at the Chicago World's Fair. Turner's reasoning went something like this: The American identity was forged where civilization abutted wildness. The tension between the two increased over time as the frontier moved, with each generation becoming more individualistic, egalitarian, mobile, nationalistic, violent, self-reliant, and dynamic. Turner saw that the frontier was "closing" as the nation reached the edge of the Pacific Ocean, and he predicted a painful adjustment to new limitations.

When you take a closer look, however, there are gaping holes in the cowboy's vaunted self-reliance. Nearly all cowboys labored for other people who owned the land and the cows. Cowboys worked for wages, without the protection of a union or even a written contract.

In *Culture of Honor: The Psychology of Violence in the South,* Richard E. Nisbett and Dov Cohen speculate on why herders are likely to be more individualistic and violent than farmers. Their theory goes something like this: Cows and sheep are easier to steal than food crops, so the herder has to be constantly on guard and able to fend off rustlers. If he is perceived as weak, he's in trouble, because he often works alone and far from others. In such an environment, arguments are personal, and a culture based on honor (as

Range showdown: Kholten Gleave, right, of Utah, pauses for the national anthem outside of Bunkerville, Nevada, while gathering with other supporters of the Bundy family to challenge the Bureau of Land Management on April 12, 2014. Photo: AP Photo/Las Vegas Review-Journal, Jason Bean.

opposed to conscience) tends to develop. Farmers, in this theory, are more dependent on each other to bring in harvests, raise barns, and the like, so they have more sense of community and less need of violence.[23]

Cowboys' golden age occurred just after the Civil War, when the United States was in the process of redefining itself. And the cowboy's Camelot, the era of the great cattle drives, lasted a mere two decades. Today America is in the early stages of redefining itself again, in a way that recognizes the finite nature of resources on a solitary blue planet. But the cowboy ideal has not yet ridden into the sunset. Today's ranchers and affiliated businesses take full advantage of this powerful myth, proudly portraying themselves as individualistic loners. A cynic might note, however, that they have been remarkably unified and effective when it comes to fighting *for* federal subsidies, but *against* federal regulation.[24]

Cowboy values also still reverberate loudly in business activities like leveraged buyouts, derivatives trading, mineral extraction, and the foreign policy of a recent president from Texas. The explosive expansion of casino gambling and lotteries suggests we haven't lost a reckless attraction to risk. And it's true that in today's world some men seem sort of lost, sitting in ergonomically designed chairs in windowless cubicles, staring at screens worked by keyboards with keys too small for their fingertips. Even if they are taking hair-raising risks with enormous sums of other people's money, this is not akin to the physical risks associated with heroic male action. Wearing cowboy boots to work lifts a guy's spirits. When a governor of Texas recently switched from cowboy boots to shoes, it made the newspapers. According to consumer researcher Russell W. Belk: "The American cowboy heroic myth is invoked by these boots, along with the characteristics associated with this myth, including rugged individualism, independence, quiet strength, and alienation from civilization."[25] A positive emotional response to cowboys is practically ingrained in most Americans. When we saw the above photo, we smiled at how well it captures what is attractive about cowboys. Our smiles faded when we learned these men were supporting a rancher who thought he was justified in running his cattle on federal land without paying a fee or obeying the law.

In thousands of ways we still pay daily tribute to the cowboy. Do you know anyone who doesn't own at least one pair of jeans? The average American woman is said to own eight.[26] Denim britches, now common in every nation on earth, were invented during the California gold rush, quickly adopted by cowhands, and worn by movie stars who played at being cowhands. A hundred years after their invention, they caught on in the 1950s with American teenagers and then with everybody from James Dean to Steve Jobs. Rappers wear them and so do women nine months pregnant. Only small details distinguish the $30 kind from the bespoke sort that cost well over $1,000. Most wearers want their jeans to look the way they look on cowboys: faded, flayed, and shrunk to fit like a second skin.

The appeal of the open range is almost entirely metaphorical, of course. Oil barons in Houston don't want cow piss on their Lucchese Belly Gator cowboy boots. Wall Street titans decked out in jeans at the annual Allen & Company retreat in Sun Valley don't dream of herding cows through mud and dust for low pay.

Although the cowboy may have been a unifying image following the Civil War—and in many ways still is—his ethos has also become a lazy way of categorizing Americans. Those who buy into the cowboy myth are more likely to watch Fox News, loathe taxes and regulations, and interpret the Second Amendment as giving them a right to pack heat when visiting Starbucks. If you're anticowboy, you're more likely to favor extending unemployment benefits, subscribe to the *New York Times*, support limits on the possession of firearms, and be less troubled by reporting to a female boss.

Dan M. Kahan, a professor of law and psychology at Yale, wrote a much-cited article suggesting that this polarization of Americans is due to "protective cognition." When it comes to evaluating scientific evidence of environmental risks like global warming, Americans tend to split into two groups: (1) those who admire equality and community (they favor abortion and same-sex marriage), versus (2) those who admire individual initiative and respect authority (they oppose abortion and same-sex marriage). Ultimately, we all hope for the same things: to be healthy, safe, and free of economic want. So why does someone's position on, say, abortion, pretty much predict his or her view on whether climate change is a threat? It seems totally random.

Kahan posits that this happens because it is almost impossible for the individual to evaluate technical and scientific data in today's specialized world. When faced with new concepts, we deal with evidence selectively and tend to trust those who we think share our values—in other words, we trust those who dress and use language like members of our own group.

One lesson from Kahan's research is this: To get a fair hearing for your ideas, it can be helpful to mentally and literally put yourself in the other guy's cowboy boots or Mephisto sandals when you are presenting your argument.[27] Denis was a jump ahead of Dan Kahan.

Denis, 1970. During the first Earth Day, Denis's attire may have helped to deflect "protective cognition." Photo: John Olson, used with permission of Getty Images.

While organizing the first Earth Day, in 1970, he stumped the country wearing cowboy boots, a cowboy hat, and a leather jacket, while arguing that tough environmental regulations were essential to protect the American Dream. When he and Gail first met, in the green room at KCET television in Los Angeles where Gail was co-host of a talk show, she took one look and assumed some incorrect things about him. It took considerable effort on his part to get her to take a second look.

Throughout this book both of us have mentally put on the garb of cow-lovers. The garb is not camouflage. We subscribe to the *New York Times*, but we also have a genuine affection for cows, an appreciation of all they add to our lives, and deep respect for many of the men and women who choose to work with them.

Cows Move Right into Our Homes

Kids learn in elementary school that meat, milk, and leather come from cows, that cows can somehow be turned into soap and candles, and that pioneers burned dried cow dung in their fires. And that was about all there was to know until the last half of the twentieth century. Then things completely changed.

Today, bits of cow are found almost everywhere. Imagine, for a moment, that every item that contains some cow emits a soft glow. It's late and you're getting ready for bed. You drop your clothes in a glowing heap (cow is used in detergents and fabric softeners). Glancing in the mirror, you see that your teeth look supernova bright (cow parts are in toothpaste and mouthwash) and your lips are shining (lipstick). Your hair's a mess, so you pick up your glowing comb (plastic) and run it through your dazzling hair (shampoos, rinses).

Under your skin, the collagen you had injected to fight wrinkles glows faintly, as does your jar of anti-aging cream. Your underarms and cheeks also emit an eerie light (deodorants and shaving cream contain cow).

You reach for a bar of soap. Made with rendered beef fat, the bar is ablaze. The sun has gone down, but you don't need to turn on lights because the walls (paint, wallpaper, wallpaper paste, Sheetrock) and floor (floor wax and linoleum) cast enough light. The adhesive that holds together your plywood cabinet is derived from cow blood.

This imaginary adventure is turning into a Stephen King nightmare. You jerk open your medicine cabinet to find your anxiety medicine, and a burst of light smacks you in the face. The gel coatings of many pills and vitamin capsules have a bovine pedigree. More light comes from bandage strips, emery boards, contact-lens-care solution, cough syrups, lozenges, and suppositories. Many medicines might contain cow material: pancreatin, rennet, estrogen, pepsin, plasmin, oxytocin, corticotrophin, fibrinolysin, glucagon, thrombin, trypsin, chymotrypsin, pegademase, cortisone, epinephrine, vitamin B_{12}, iron, thrombin, chondroitin sulfate (made from a cow's nasal septum) and deoxyribonuclease are examples. Vaccines (a word coined from the Latin word *vacca*, cow) may be prepared using a medium that contains cow. Medical syringes and surgical sutures often incorporate cow.

Dazed, you leave the bathroom and wander through your house. Bone china, violin strings, ice cream, jelly beans, gummy bears, mayonnaise, Jell-O, piano keys, chewing gum, marshmallows, sweeteners, printing ink, high-gloss finishes for magazine paper, glue in books, cleaners, candles, matchstick heads, pet foods, textiles, buttons, upholstery, insecticides, glues, crayons, and the film inside that old Kodak camera you've been meaning to give to Goodwill all radiate light.

You peek into your garage. There's some cow in antifreeze, tires, upholstery, hydraulic brake fluid, car polishes and waxes, asphalt, bonemeal fertilizer, industrial cleaners, baseballs, machine oils, auto lubricants, paintbrushes, and foam for fire extinguishers (made from protein recovered from cows' horns and hooves).

What we use from cows, and how we use it, constantly changes. The same smart chemists who figured out how to extract useful molecules from cows have also figured out how to make synthetic versions of many of these things. Take the field of medicine: A website run by the state of Michigan says: "One of the most important cattle by-products is insulin, a treatment for people with diabetes."[28] This was once correct. But today human insulin is made using recombinant DNA technology. Another medicine, thyroxine, was formerly made from dried cow and pig thyroid glands. Today a synthetic thyroid hormone, Synthroid, is used by nearly everyone. The anticoagulant heparin was originally made from the lungs and intestines of cows and pigs, but it, too, can now be synthesized. The same is true of epinephrine (adrenaline), used to revive people suffering from asthma or shock, which once came from cows' adrenal glands. Synthetics are also replacing surgical sutures once made from cows' intestines.

Moving on to the world of sports: "Catgut" tennis racket strings (actually made from cows' intestines) are still around and used by some pros. But most racket strings today are made from the more durable Kevlar, nylon, and polyester. And although as many as twelve basketballs can be made from a single cowhide, NCAA rules now also allow a ball to be covered with either leather or composite material—as long as the ball is orange, red-orange, or brown.

Sometimes synthetics are cheaper and environmentally friendlier, sometimes not. The point here is that they exist. So do we still need cows to supply us with products other than milk and meat? Clever humans will probably never stop thinking up new uses for cow parts. Some of these potential uses sounded like science fiction to us until we took a closer look. For example, cow brains might lead to smarter energy sources. Huh? It turns out that cow brains contain a multitalented protein called clathrin, which, given the right encouragement, can form self-assembling shapes such as cubes, spheres, and tetrahedra. Clathrin can do this without nasty chemicals or the use of high pressures and temperatures. Scientists at Stanford University are adding atoms to clathrin structures to create things on a nanoscale that might someday make excellent batteries and solar cells.[29]

After six years of effort by an international team of hundreds of scientists, the cow genome was unraveled and published in its entirety in 2009.[30] About 80 percent of cows' 22,000 genes are also found in humans.[31] This genetic understanding is leading to all kinds of things. Scientists have already inserted a handful of human genes into cow embryos to get cows to produce substances not normally found in cow milk or blood. In cow milk this includes human growth hormone,[32] skim milk straight from the cow,[33] and milk that's more like human breast milk than regular cow milk. Cow blood from genetically engineered cows may soon contain human polyclonal antibodies to fight infection, inflammation, cancer, and autoimmune diseases.[34]

Decoding the cow genome has also led to what may be a major breakthrough in the understanding of evolution. It seems that cows and snakes share up to a quarter of their DNA, something called BovB, thanks to . . . blood-sucking ticks? That's what some scientists think. If correct, this suggests that genes can be transferred not just from parent to child, or between individuals by retroviruses, but between different animal species, with a third species acting as carrier! Elephants and platypuses were also found to have the same repetitive stretches of genetic coding.[35]

Most Americans would consider it ethical to use part of a cow to save a human life. But it's easy to imagine some cow/human blends that would cross the line for many, such as cows with recognizably human features or traits. As is often the case, science is outpacing our legal system. For example, how many human genes, and which ones, should it be legal to insert into a cow genome? Should a cow carrying many parts of the human genome have a human guardian appointed? Would the cow get standing in court and have legal rights? Is the welfare of cows adequately protected throughout such experiments? Human DNA has already been inserted into scooped-out cow eggs and induced to grow by electrical shocks. The embryos were 0.1 percent cow and 99.9 percent human and were not allowed to grow beyond three days.[36] But what if cows were used to carry human babies all the way to birth? Or (if it became possible) to birth a creature that is, say, 5 percent cow and 95 percent human? Even many enthusiastic supporters of genetic engineering are troubled by the notion of part-human chimeras.

And then there is also the intractable problem of social awkwardness: "Son, I'd like you to meet Clover, your birth mom."

Chapter Two

WHOA! COWS' BIG BITE

We sat facing each other on boulders overlooking Icicle Creek in the Cascade Mountains. The creek rumpled as it bounced over rocks. Fat with melting snow, it spit spray into the air. Not a cow was in sight, but we talked about the book we were writing. The conversation turned to cowboys, about how cowboys were our heroes when we were young, and how the image of independent loner might have been a catalyst for Denis's long, solo hitchhiking trip around the world when he was nineteen. He rented surfboards in Hawaii, taught swimming in Tokyo, rode the rails across what was then the Soviet Union from Vladivostok to Moscow in winter, and wandered his way around Africa.

Today, however, "loner" calls to mind the nut up in a tower with a rifle, or the guy in the woods wiring bombs and penning a manifesto. As Frederick Jackson Turner predicted, the frontier is gone. From where we sat, however, there was no sign of civilization. A salmon arced out of white water, disappearing immediately into the big gush. Ponderosa pines leaked a vanilla tang into the air. Across the creek, pyramids of pine covered the mountain.

The suggestion of unclaimed possibilities was deceptive, however. Every inch of this wild landscape was either privately owned or federally protected. And downstream owners of water rights had appropriated every drop of water in Icicle Creek. In fact, more water had been allocated than was actually in the stream.

Sign at entrance to ranch off Northern Cascades Highway, Washington. Photo: Gail Boyer Hayes.

Gail is the daughter of a biochemist and was raised during the optimistic Better-Living-Through-Chemistry era in the 1950s. Her young life was one of rising affluence. She tends to assume that, in this era of great scientific and technological advances, most problems, including cow problems, can be solved.

Denis was raised in paper-mill towns where residents called toxic, acidic air "the smell of prosperity," where his dad worked in the mill and had to wash his car every day if he didn't want the paint eaten off. The mill had a free drive-through car wash that anyone in town could use. But there was no similar protection for people; Denis woke up with a sore throat every morning for the first eighteen years of his life. He admires the creativity of entrepreneurs but has no illusions about the ruthless single-mindedness of those competing for market share.

We both suspected from the start that cows currently demand too big a share of the nation's resource pie for the benefits they provide. But we didn't realize how very large and destructive that share has become.

Cows and Climate Disruption

Overall, America has a fine temperate climate that serves us well and that should be counted as one of our important natural resources. The warming of Earth's atmosphere and oceans has begun to make that climate less benign.

Feed production and processing, the fermentation that goes on inside cows, and the decomposition of cow manure release three warming gases: carbon dioxide, methane, and nitrous oxide.[1] A report issued by the Food and Agriculture Organization of the United Nations in 2013 concluded that livestock account for 14.5 percent of anthropogenic (human-caused) greenhouse gas emissions. Beef cattle are responsible for 41 percent of livestock emissions, milk cows for another 19 percent. It would be prudent to reduce those emissions.

Since the start of the Industrial Revolution, human activity has increased the amount of carbon dioxide (CO_2) in the atmosphere by 40 percent. University of Chicago oceanographer David Archer warns, in *The Long Thaw*, that humans are putting carbon dioxide into the air that could stay there for centuries, *a quarter of it lasting essentially forever.* Archer advises, "The next time you fill your tank, reflect upon this."[2] Think of it, too, the next time you bite into a cheeseburger. Because many aspects of climate change are now inevitable, scientists and planners have recently replaced the word "sustainability" with the word "resiliency" when speaking of their goals for many communities. Although still hoping to reduce even greater warming, they are also looking for ways for communities to adapt to those shocks now certain to come. (Unless, of course, something strange and unforeseeable happens, such as the Yellowstone mega-volcano awakens or a big asteroid hits and kicks up a worldwide dust storm. If that happens we'll have more pressing concerns.)

Globally, the livestock sector accounts for 9 percent of human-caused warming that's due to increases in the gas carbon dioxide.[3] Most of this comes indirectly, such as from cutting down forests to create pastures and growing feed for cattle. Carbon dioxide is also released during the manufacture of the fertilizers and pesticides that drench the crops used to feed confined cows. Burning fossil fuels to power farm machinery, and to transport cows, meat, feed, water, and cow waste, also creates this gas.

Methane (CH_4) is another powerful warming gas, one that hasn't received the public spotlight it deserves. Although it doesn't stay in the atmosphere nearly as long as carbon dioxide does, methane is far better at trapping heat. So over the course of a century, methane contributes more to warming than does carbon dioxide. Human activities have put more methane into the atmosphere today than it has contained for at least four hundred thousand years, and those emissions are accelerating worldwide.[4] Globally, around 14.5 percent of all anthropogenic warming emissions are due to this gas,[5] and around 37 percent of anthropogenic methane comes from livestock, mostly from cows.[6] Methane is produced when cow manure breaks down. It also comes from cow burps (and a bit from cow farts).

A report issued in December 2013 by the National Academy of Sciences found that livestock operations in the States emitted twice as much methane in 2007 and 2008 as previously believed.[7] The new study included measurements from actual samples of air, whereas earlier studies by the Environmental Protection Agency (EPA) and an international group had calculated emissions based on assumptions and models.

Methane emissions can be reduced by feeding cows higher-quality (and more expensive) forages such as flax and alfalfa seeds,[8] or by providing cows with more fat, particular proteins, tannins, nutrient-laden salt licks, certain medications, fish oils, or a new "burpless grass."[9] Canadian scientists came up with a diet for beef cows that balances starch, fat, sugar, cellulose, etc., and reportedly cuts methane emissions by a quarter.[10] Cows will eat an astonishing variety of biowaste. Some are obligingly swallowing a weird array of foods like cashew-nut-shell liquid and whole cottonseed to see if doing so reduces their methane output. A by-product of winemaking, grape marc, reduces dairy-cow methane emissions by 20 percent.[11] Cows don't normally eat their own bedding—but if the hay and straw are cut into pieces about three inches in length and tossed with a few goodies, cows will consume it. Milk yield goes up and methane burps are 20 percent fewer.[12] A Penn State dairy scientist found our favorite partial solution: feeding cows the culinary herb oregano. This cut methane emissions by 40 percent, boosted milk production, and gave Clover the freshest breath in the barn.[13]

Farmers won't adopt new feeds unless they are also affordable and reasonably easy to implement. Moreover, even with all these dietary tweaks, cows will still release massive amounts of methane.

Finally, cows' manure releases the warming gas nitrous oxide (N_2O). Your dentist may have offered you nitrous oxide before a painful procedure. "Laughing gas" induces a feeling of euphoria and brings on a dreamy mental state. Molecule for molecule, nitrous oxide is 310 times more potent than carbon dioxide as a greenhouse gas, and it has an impressive atmospheric lifetime of around 120 years. The livestock sector is responsible for 65 percent of the anthropogenic nitrous oxide in the air, most of which comes from the breakdown of cow manure.[14]

Just as cows have a big effect on climate, climate has a big effect on cows. The eight hundred existing breeds can handle a wide range of climatic conditions as long as they have water to drink, food to eat, and a veterinarian with a charged cell phone. However, our enormous national herd is largely composed of just a handful of breeds that have been optimized not for survival but for producing great quantities of milk or marbled beef while living in the climate that prevailed until recently.

When cows get too hot, they lose their appetites and fertility. They suffer even more when the humidity rises along with the air temperature. At 92 degrees Fahrenheit and 100 percent relative humidity, cows start dying. Stressed cows need higher-energy-value feed and much more water. Both feed and water are usually in short supply when the heat coincides with drought, as is often the case.

American cows are already suffering from climate disruption. In October 2013, tens of thousands died when an early blizzard hit South Dakota; the cows hadn't had time to grow their thick winter coats. Over three recent years, more than a million cows were lost to the worst drought in Texas's history. *Los Angeles Times* reporter Julie Cart wrote: "[M]ore big blows than anyone can remember have roared through this year, stripping away precious topsoil and carrying off another season of hope for farmers and ranchers." An unprecedented number of wildfires burned thousands of square miles and thousands of homes.[15] Exceptional drought (the most severe category) returned in 2012 and yet again in September 2013.

What we've seen thus far has been milder than what's coming. The United States is expected to experience more violent weather—

more flooding, windstorms, hurricanes, hail, dry stretches, cold snaps, and hot spells. Crop failures will mean less food is available for cows. Frequent hot spells will require farmers to provide more air-conditioning for dairy cows, more shade for beef cattle, and more water for both.

Magical Thinking About Rain and Water

Earth might be the blue planet, but only 1 percent of its water is in a form humans can use. And Americans are clamoring for more fresh water every year. Thirty-six states expect imminent water shortages, and almost every region has already had to grapple with shortages.[16] To date, drought has affected more Americans than any other natural disaster.

Water is needed at every stage of meat and dairy production: for cows to drink, to grow feed crops, to clean dairies, to wash cows before milking or slaughter, to wash the person who milks or processes them, to wash carcasses, and to clean slaughterhouses. Vaclav Smil, professor emeritus at the University of Manitoba, notes that "few economic endeavors are as water-intensive as meat production in general and cattle feeding in particular."[17] According to Sightline Institute president Alan Durning, it takes 840 gallons of water to create one pound of grain-fed beef.[18]

As for dairy, professor Jonathan Foley, director of the Institute on the Environment at the University of Minnesota, sums it up colorfully: One Starbucks Frappuccino requires three hundred gallons of water, and the vast majority of those gallons go into making the cream.[19] Estimates of how much drinking water a lactating dairy cow needs vary by state (and by estimator), but everyone agrees she consumes more fresh water than nearly any other animal. An Ohio study estimates about thirty gallons per cow per day are needed for drinking and cleanup.[20] In hot, dry Texas, the average is over sixty-eight gallons per cow for drinking and sanitation. Slaughtering reportedly uses around 132 gallons per animal carcass.[21]

Over half the U.S. sales of cattle and calves originate in one of five states, and four of those states are on the Great Plains: Texas, Nebraska, Oklahoma, and South Dakota. Most of the food fed to confined cows is grown on the plains.[22] Surface water is scarce over the portion known as the High Plains, which is why for thousands of years there were no permanent settlements there. In 1862, however, when President Lincoln signed the Homestead Act, the head of a household could claim 160 acres for $18 and five years of plowing. Newly established railroads also sold settlers much of the land Congress had granted them (a subsidy of ten square miles for every mile of track laid). More settlers meant more rail customers.

The Union Pacific touted the "unsurpassed grazing," "mild and pleasant" climate, and "pure and abundant water" in Kansas. After they arrived, settlers confronted tornadoes, which the ads had neglected to mention. "Pure, abundant water" also reflected great poetic license. However, most settlers believed that "rain will follow the plow." Plowing would expose more moisture in the soil to the air, they reasoned, and that moisture would turn into rain. Well, moisture went up into the air all right, but it joined the great air currents and fell far from the plains. Settlers hung on by allowing themselves to believe some fantastical things: that smoke from trains, metal in rails, or vibrations from human activity would soon bring rain. When no rain came, they tried dynamiting the air.

Homesteaders and their oxen plowed up the prairies. Moisture-sucking-and-transpiring annual crops like wheat and corn replaced deep-rooted perennial native grasses. The result was the Dust Bowl of the 1930s, when topsoil built up over thousands of years blew away. Ten million acres lost five inches, and another fourteen million acres lost two and a half inches.[23] In many areas, that was more than three-fourths of the topsoil. The Dust Bowl was a traumatic event in American history and has been immortalized in fiction by John Steinbeck and in nonfiction by Tim Egan.[24] But few Americans recall that another devastating drought occurred on the plains in the 1950s, when some ranchers survived by feeding their cows prickly pear cactus mixed with molasses. In the late 1980s, yet another Great Plains drought cost the nation more than any other natural disaster up to that time.[25]

And here's the rest of the story: Paleoclimatic data (ice cores, rocks, sediments, tree rings, fossilized bison teeth,[26] microfossils, etc.) indicate far greater climate variability in North America than what is revealed by relatively recent instrumental data. Perhaps due to solar cycles, over the last 4,500 years the climate on the northern part of the Great Plains has gone through drought/wet cycles roughly every 160 years. Some droughts have lasted several decades and affected areas now in the Dakotas, western Minnesota, eastern Montana, Wyoming, and the adjacent parts of Canada. Much of this land is now used for grazing cows.[27]

The worst drought of the past 1,200 years happened in the mid-twelfth century, lasted sixty years, and was due to natural climate variability.[28] It covered most of the West (including the Pacific Northwest), the Great Plains, and northern Mexico. There is no reason to believe such severe drought can't happen again. Humans can't prevent this, but we could mitigate some of its effects.

• • •

Early in the twentieth century a great buried treasure was discovered: the vast pool of fossil water called the Ogallala Aquifer. "Pool" is somewhat misleading: think waterlogged gravel, sand, and clay lying deep underground. The Ogallala is one of Earth's largest aquifers. It lies beneath Nebraska and parts of Texas, Kansas, Oklahoma, Colorado, Wyoming, New Mexico, and South Dakota. If it weren't for the discovery of the Ogallala, farming on the plains would never have recovered after the Dust Bowl.

The thickness of the saturated layer varies from more than a thousand feet under parts of lucky Nebraska to only a few feet elsewhere. If the water originally in the Ogallala were spread across the entire United States, it would form a layer a foot and a half deep. Over the last century, however, we've pumped out two-thirds of the total water, enough to fill Lake Erie.

The invention of center-pivot irrigation after World War II made it possible to fully exploit this resource and turn the region into the nation's breadbasket. When you fly above the Ogallala, you see that the land below you is covered with giant green polka dots formed by irrigation systems that tap into the aquifer and pivot from a central point. The sprinkler arms are half a mile long, reaching from the pivot point to the edges of one-square-mile sections of land. Revolving over the field like the hands of gigantic clocks, the units cost $180,000 each.

Although few coast-dwelling Americans even know the Ogallala exists, it is the source of nearly all the water used in the High Plains. In fact, it supplies nearly a third of all the groundwater the entire nation uses for irrigation. *A fifth of all the agricultural products in the United States and a full 40 percent of the grain for grain-fed beef are grown with Ogallala water.*[29] Corn (cow food) is the most profitable crop grown above the aquifer, and also the thirstiest.

The fossil water in the Ogallala is not renewable.[30] Over the years, water levels have fallen by more than one hundred feet in parts of

Pivoting irrigation systems create circles of crops in Kansas, and suck water from the Ogallala Aquifer. Shortgrass prairie once covered this region. Photo: NASA/GSFC/METI/ERSDAC/JAROS, and U.S./Japan ASTER Science Team.

four states. The "tragedy of the commons" is writ large over the Ogallala. Like climate disruption, aquifer depletion is the type of problem that human minds aren't well designed to handle: The problem spans generations, conditions are only gradually worsening, and most people find it in their short-term interest to behave in ways that benefit them but harm future generations. Underground water ignores property lines, so there's no way to conserve the water under your land unless all your neighbors do the same. Therefore, it seems logical to pump out as much water as hard and fast as you can, because otherwise your neighbors will pump it out. Many geologists expect most of the Ogallala to run out of water in twenty-five to thirty years, and perhaps run out of "usable" water as soon as 2020.

Management of this invaluable resource is inconsistent, complex, and scattershot. Decades of withdrawals plus three years of drought have greatly reduced the amount of water that can be pumped from the southern two-thirds. Many formerly green circles have gone brown or are now only half circles. They won't come back. Large parts of southern Kansas, the Oklahoma panhandle, and northwest Texas are already in crisis, and smart farmers are switching from corn to sorghum and cotton in an effort to make the water last longer.[31]

Most states above the Ogallala impose a dollop of regulation through the issuance of new water permits. In Kansas, for example, pumps are metered and withdrawals limited. But even in Kansas many old water rights have been grandfathered, so that the heirs of rights holders or legal entities that first put a certain amount of water to a beneficial use continue to have that right.[32]

Texans can pump all the water they want from beneath their land. Although Texas pioneered the first legal mechanisms to protect mineral owners from having their oil poached by neighbors, the state steadfastly clings to a nineteenth-century view of water allocation. Until recently, billionaire T. Boone Pickens's Mesa Water company owned the groundwater rights to more than two hundred thousand acres in Roberts County in the north Texas Panhandle. Those rights made Pickens the largest private water owner in the nation.[33] Pickens tried unsuccessfully to sell his water to distant cities like Dallas and San Antonio. To get the necessary eminent domain to run a pipeline and power lines, he created his own town, a town with only two eligible voters. But after a decade of playing footsie with distant cities, he finally sold for less profit than he'd hoped for, to an outfit that will market the water to desperate cities nearby in the Panhandle.[34]

The aquifer's water was once of such fine quality that you could drink it unfiltered and untreated. Now, however, the EPA says that pesticides, fertilizers, feedlot wastes, trace metals, and volatile organic compounds have contaminated much of it. In researching this chapter,

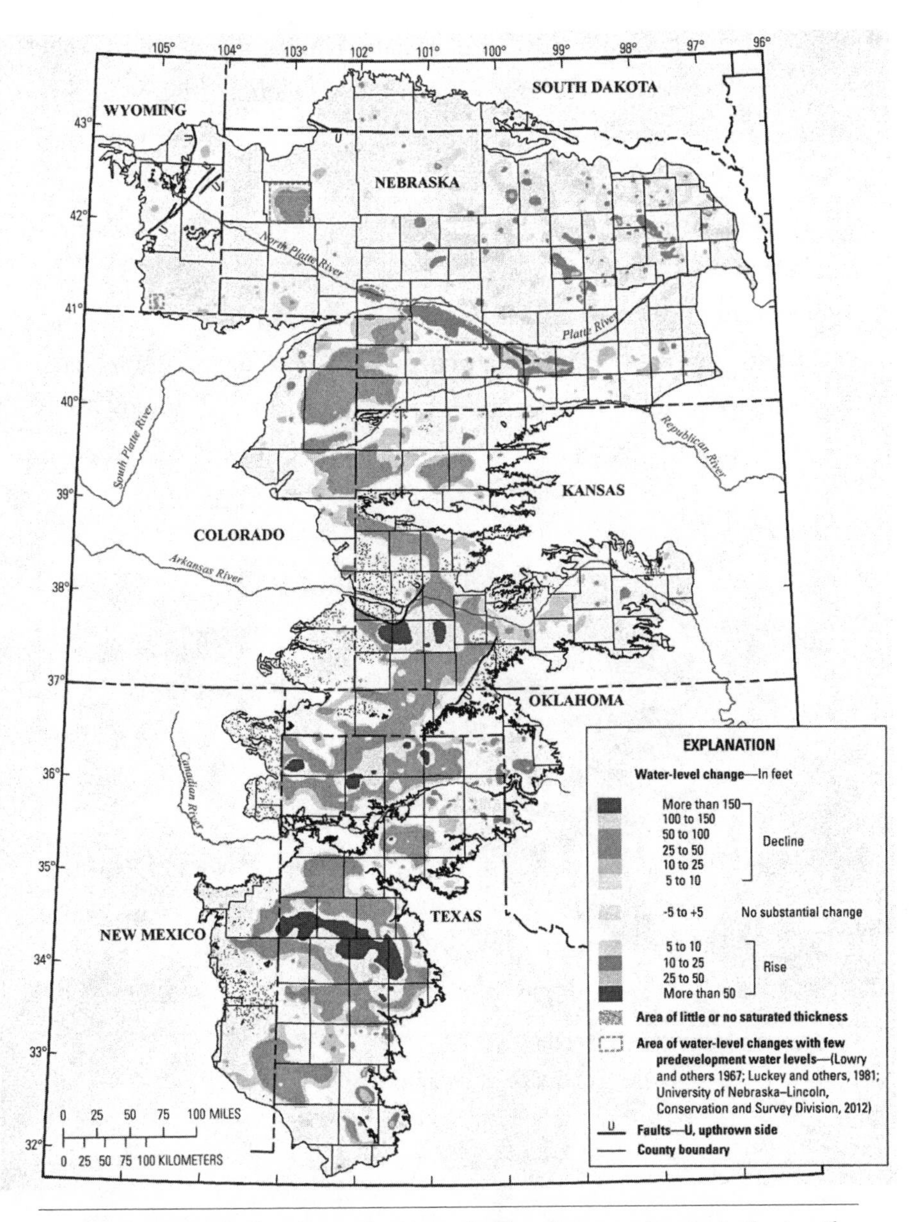

The Ogallala (High Plains) Aquifer. Water-level and storage changes in the Aquifer predevelopment to 2011. This fossil water irrigates the U.S. breadbasket and is being depleted at an alarming rate. V. L. McGuire, "Water-Level and Storage Changes in the High Plains Aquifer, Predevelopment to 2011 and 2009–11: U.S. Geological Survey Scientific Investigations Report 2012–5291," 2013, 15 pp. Also available at http://pubs.usgs.gov/sir/2012/5291/. Figure 1, p. 2.

we stumbled across a disturbing correlation. We noticed a remarkable similarity between a map showing where concentrated cow-feeding operations are densest and maps of the Ogallala. The entire surface above the precious aquifer is dotted with polluting cattle feedlots.[35]

Dave Brauer, a research agronomist with the Ogallala Research Service of the United States Department of Agriculture (USDA), told reporter Charles Laurence of the *Telegraph* (UK) that he thinks it's too late to save farming on the plains. The USDA's goal, Brauer said, is just to slow down the inevitable and mitigate the damage.[36] There

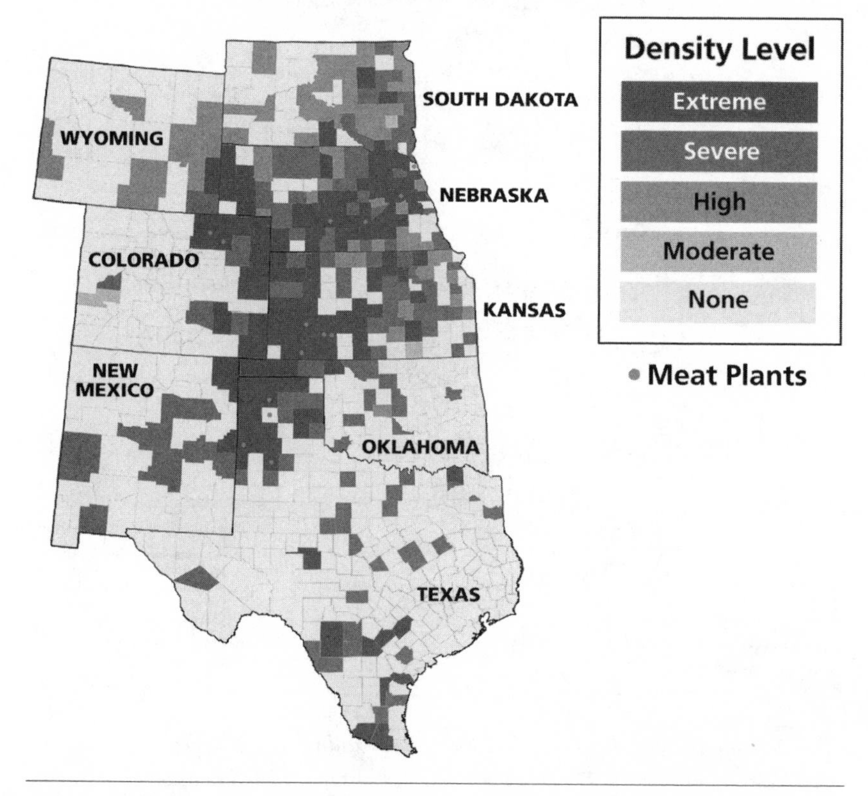

Density of cattle factory farms, 2007. Most cattle factory farms are located above the Ogallala Aquifer, draw water from it, and potentially pollute this valuable resource. Image and data courtesy of Food & Water Watch, http://foodandwaterwatch.org, and factoryfarmmap.org.

is speculation about undertaking massive geoengineering solutions, seawater desalination, and so forth, but they are not serious proposals from knowledgeable people with financial backing. No one has suggested any source of remotely affordable water to irrigate the plains when the Ogallala is drained.

Skinning Our Land

Earth is unique among known space rocks in that it has a thin skin of living dirt. Without dirt there would be neither cows nor humans. Earth has three major regions of fertile loess soils: our Midwest, northern Europe, and northern China. (Loess is easily crumbled buff to brown dirt made of clay, silt, and sand.) These regions provide the bulk of the world's grain. On the Great Plains, it takes Mother Nature five hundred years to replace one lost inch of loess soil. In North America, 66 percent of soil loss comes from agricultural activities and another 30 percent from overgrazing.[37] "Every second, North America's largest river carries another dump truck's load of topsoil to the Caribbean," writes David R. Montgomery, the winner of a MacArthur "genius" grant and a geomorphologist. He argues convincingly that we are skinning America and warns that we could be undoing the very ground that made us a wealthy nation.[38]

More than half of America's cropland has a slope of at least 2 percent and is thus vulnerable to water erosion when it is plowed.[39] In an article for *Harper's,* Richard Manning wrote: "[I]f you can find what Iowans call a 'postage stamp' remnant of some [prairie] it most likely will abut a cornfield. . . . Walk from the prairie to the field, and you will probably step down about six feet, as if the land had been stolen from beneath you."[40]

There are many ways to conserve soil, including cover crops, buffer zones alongside waterways, contour plowing, crop rotation, no-till planting, mulching, and terracing. Congress occasionally passes a law to reward conservation that enjoys a fleeting success. From 1985 until 1997, because of the "highly erodible land conservation" rules in the 1985 Food Security Act, soil erosion was reduced by 40 percent. But in 1996, under pressure from Big Ag, Congress greatly weakened those provisions. They did so because soil conservation techniques take marginal land out of production, preventing farmers from maximizing their current income. Because of federal subsidies and the mandated use of ethanol, twenty-five million additional acres have been plowed in the United States since 2007.[41] At the same time, the Midwest was hit with high-intensity, high-volume rainstorms. The value of the soil washed away was surely greater than the value of the extra grain grown.

The most recent farm bill slashed funding for conservation programs by $6 billion, but did make adopting a conservation plan a requirement for getting crop insurance subsidies. It also included a mixture of other wins and losses for conservation advocates.[42]

Iowa grows more corn for cows than any other state. So let's look at what's happening to Iowa's dirt.[43] Over the last 150 years, Iowa has lost half its topsoil. Iowa State University scientists working on the Iowa Daily Erosion Project say erosion is worsening and that much more soil is being lost than is acknowledged in the official estimates of the USDA's Natural Resources Conservation Service (NRCS).[44] In 2007, more than ten million acres of Iowa farmland eroded faster than the NRCS's so-called "sustainable" level of five tons per acre per year. Six million acres lost soil at twice the sustainable rate.[45]

Most farmland in Iowa is now leased.[46] When old farmers retire or die, their children often want to hang on to profitable land but don't want to be farmers themselves. So they rent it out. Renters don't know the land as well as long-term owners, nor do they have much incentive to take care of it. Money that a renter invests in long-term improvement will benefit the next person the owner rents to, not the current renter. His sole goal is to grow as many bushels of corn as he can. Sadly, the regions in Iowa with the most fertile soil tend to be those where the highest percentage of farmland is leased. In north central and northwest Iowa, as much as 70 percent of the acreage is farmed by someone who doesn't own it.

Iowa State University Extension sociologist J. Gordon Arbuckle Jr. looked into the implications of leased farmland. He discovered that landlord-tenant relationships and landlords' conservation ethics both deteriorate with distance. Four out of ten tenants polled thought conserving the soil and water quality was the landowner's responsibility, not theirs. Legally, they were probably right.[47] It would be wise to change the law to impose responsibility on renters as well as owners, perhaps by mandating such clauses in lease contracts.

The quality of soil matters as much as the quantity. A single handful of dirt swarms with billions of microorganisms (bacteria, fungi, protozoa, nematodes, mites, microarthropods) that can be harmed by many things. We have little idea what we're doing to them as we douse them with synthetic fertilizers, pesticides, and herbicides. Furthermore, repeated tilling compacts what was once granular, spongy prairie topsoil, so that water and roots can't move though it as easily.[48]

Holistic Management

Cornell University researchers estimate that it would cost $44 billion a year to undo the damage we're doing to our soil but would cost only about $8.4 billion a year to bring soil creation into balance with soil erosion on crop and pasturelands. So investing $1 could save $5.24.[49] With so much of our soil already degraded, however, soil also needs to be rebuilt. Rebuilding good soil can also help clean our air and water.

We found a great many innovative approaches—some of them already widely practiced by small farmers and ranchers—that allow cows not merely to cause less harm but actually to benefit the land. The size and scope of the benefits is a matter of some debate, but the possibility of beneficial effects is clear. One approach that kept coming up again and again is holistic management.

There's an almost legendary TED talk by Allan Savory, originator of holistic management, titled "How to Green the World's Deserts and Reverse Climate Change." The talk has been viewed more than three million times. More than any other vehicle, that talk propelled holistic management into the mainstream and turned Savory into an environmental rock star.

Allan Savory is a bigger-than-life character who, in his youth, was an important political leader of the moderate forces in Northern Rhodesia. A biologist by training, he initially shared the view, still held by most, that overgrazing was destroying the soil. In the 1950s, he participated in the culling of twenty thousand elephants, something he now terms the "saddest and greatest blunder of my life."[50] Today, Savory argues passionately and persuasively that heavy grazing—coupled with frequent, rigorously controlled rotation among paddocks—can actually *benefit* land, and his followers have demonstrated this on thousands of ranches.

The underlying theory is that wild herbivores, needing protection against predators, evolved to bunch closely together. They graze intensely and then move on, prompted either by a search for new food or by predators lurking nearby. After the bison, wildebeests, or zebras move on, the intensely grazed area is left undisturbed to recover until the next herd arrives.

Savory's insight was that it would be smart for farmers to mimic this pattern. By keeping cows tightly bunched, moving them regularly from one paddock to the next, and allowing each grazed area ample time for recovery before returning, farmers and ranchers could use cattle to enrich their soil, and even sometimes restore ruined land to fertility.

In places with appropriate climate, forage, and soil, holistic management has clearly made the land healthier than it would have been if no ruminants at all lived on it, and also healthier than if it had been either lightly grazed or overgrazed.[51] The concentrated dung and urine left behind fertilized new plant growth. Trampled grasses decayed faster, allowing more sunlight to reach new growth. Soil crusts were broken so seeds and water could penetrate (a technique that works better on some soils than others).

Virtually all of the exceptional ranchers we interviewed for this book acknowledge their indebtedness to Savory's teachings. This is true despite the fact that we did not seek out holistic ranchers; we sought successful, sustainable, organic ranchers who produced grass-finished beef. Only later did we learn of their embrace of holistic methods.

So what is the controversy? Actually there are two objections, one minor and one major. Neither detracts from our support for holistic management where it makes sense. The minor one lies simply in the imperfections of human beings. Savory's message has nuances that sometimes are lost as his gospel is passed on through repetition. Successful holistic management, with scheduled rotations, is labor-intensive. Forage consumption and range conditions must be closely monitored, records kept, fences frequently repositioned. Cattle must be moved frequently and not allowed to roam. The land must be left undisturbed for long periods between grazing.

Unlike the rote formulas taught in agricultural extension programs, holistic management requires real-time intellectual engagement. Depending on the plants, the soils, the weather, and the breed of cattle, exactly the same vocabulary can stand for taking different actions at different times. For example, ranchers have to be alert to how fast various plants are growing versus how fast cattle are eating.

Savory is a very smart guy, and the ranches he personally manages tend to be very successful. So are those of his smart, self-starting, entrepreneurial acolytes possessing years of experience, actively engaged minds, and a personal ethic of hard work. However, that is not a perfect description of every rancher in the world. Even some ranches that flourished under Savory declined abruptly after his departure. Not every ranch that adopts the holistic vocabulary will produce exceptional results.

The major controversy has little to do with the effectiveness of holistic management on right-sized ranches and dairies in appropriate locations. Instead, it traces back to that TED talk.

The title of that talk, remember, was "How to Green the World's Deserts and Reverse Climate Change." Had the title been "Smart Rotational Grazing Can Produce Healthier Cattle While Enriching the Soil," the major controversy would not exist. However, a talk with that thrust would not have produced the viral ecstasy aroused by the prospect of rolling back deserts and curing global warming by serving everyone steak. As Savory put it in the talk: "If we do what I'm showing you here, we can take enough carbon out of the atmosphere and safely store it in the grassland soils for thousands of years. And if we just do that on about half the world's grasslands that I've shown you, we can take us back to pre-industrial levels, while feeding people. I can think of almost nothing that offers more hope for our planet, for your children and their children, and all of humanity."

That is the sort of claim that gets attention.

Some global warming deniers (which Savory emphatically is not) were, of course, delighted. (They cite the speech to show how trivial it would be to cure global warming, if warming in fact existed.) But the "let's turn the deserts into pastures and solve global warming" idea was more skeptically received elsewhere.[52]

While holistic management has been successful in enhancing pastures and restoring rangelands in temperate climates, evidence that it can roll back the world's deserts is much more sketchy. Savory seems to draw largely on his personal experience in the Charter Grazing Trials in 1969–1975 in what is now Zimbabwe. Extrapolating from seven years on a 6,200-acre semiarid patch in southern Africa (less than ten square miles) to six billion acres of desert around the world—the Sahara, the Gobi, the Namib—requires a leap of faith. Moreover, it turns out that rainfall during the Carter Grazing Trials was 24 percent above average, and on three of these seven years the land received more rain than Seattle. It is not an easy extrapolation to the Sahara.[53]

Many of us believe that it is not possible, nor desirable, to eliminate deserts. The world had vast stretches of desert before humans began influencing desertification, and indeed before humans existed. Many deserts have their own rich, complex ecosystems, and their own fierce champions. Deserts have important effects on global warming, with an albedo that reflects much solar radiation; covering them with grasses and cows would absorb far more heat.

But beyond all that, the numbers just don't add up. Because we humans have burned so much fossil fuel, cut down so many forests, drained so many peat bogs, raised so many livestock, and plowed up so many grasslands, we have increased atmospheric CO_2 by 40 percent. There is no credible livestock grazing strategy that, by itself, can begin to sequester all the world's contemporary carbon dioxide emissions in real time, much less undo the past 250 years of atmospheric accumulation.

Good grazing practices can, however, make a noticeable dent in this accumulation. When grasslands are converted to row crops, soil carbon decreases deeply and abruptly. If erosion is curtailed and smart land use policies (including rotational grazing) are embraced, much of that lost soil carbon can be restored.

Rattan Lal is Director of the Carbon Management and Sequestration Center at Ohio State University and past president of the World Association of Soil and Water Conservation. A distinguished scientist—and one of the field's outspoken optimists—he thinks that, over a period of forty years, excellent agricultural practices might restore American Midwest soils to two-thirds of their original carbon content. Moreover, Professor Lal believes that managing all the world's soil for carbon could draw down atmospheric CO_2 by 75 parts per million, or even more, over a century and a half. Since total atmospheric CO_2 is currently about 400 parts per million, this is a stunning estimate. A 75 parts-per-million reduction (coupled with no new emissions) would return us to 1970 levels.

Some people have confused the management of soil for carbon content with the adoption of extremely widespread holistic grazing. But Lal's goal is vastly more ambitious. He is talking about a heroic mobilization of the entire planet to manage all the world's soil for carbon sequestration. This would include genetically modifying cover crops and plants, using nanostructured nutrient delivery systems, superb new watering technologies, controlling desertification, enhancing wetlands, reforestation, establishing new forest, using land for the simultaneous growing of trees and crops, using biochar, and so forth.[54] Good grazing and grassland management is certainly part of it, but a relatively small part.

Even at this outer boundary of technological optimism, the ability of soil to sequester carbon rapidly runs into a ceiling. The total amount of carbon stored in soil is unlikely to return to the high levels that existed before the agricultural revolution. Unless coal use is phased out swiftly, and all fossil fuel use is terminated within several more decades, the planet's soil will have done everything it can and the growth of greenhouse gases will continue unabated.

Smart grazing is a very good thing for soil carbon, but it is just one arrow in a very large quiver of climate solutions, and not the most important arrow at that. The lesson drawn by many from Savory's talk—the more cows the better—is just dead wrong.

Holistic management is a form of husbandry for ranchers who love the land and hope to pass it on in better condition than they found it in. It is a form of ranching that requires hard work and long hours, intelligence, careful observation, and adaptability. Along with profits and healthier land, this sensitive biomimicry can yield a deep sense of satisfaction.

In 2003, a small group of ranchers, environmentalists, and scientists met to see whether they could find common ground. Exhausted by decades of conflict and litigation, they hoped to find a way to end the acrimony. To their surprise, they discovered that during those years of warfare, each had come to have a greater appreciation of the other's point of view. The ranchers had almost all drifted into rotational grazing and other holistic practices. The environmentalists, after watching species after species disappear despite their lawsuits, kept encountering small ranchers who were voluntarily managing for biodiversity. In the end, the group resolved to create the "Quivira Coalition," and they forged a moving declaration:

> We believe that how we inhabit and use the West today will determine the West we pass on to our children tomorrow; that preserving the biological diversity of working landscapes requires active

> stewardship; and that under current conditions the stewards of those lands are compensated for only a fraction of the values their stewardship provides.
>
> We know that poor management has damaged land in the past and in some areas continues to do so, but we also believe appropriate ranching practices can restore land to health. We believe that some lands should not be grazed by livestock, but also that much of the West can be grazed in an ecologically sound manner. We know that management practices have changed in recent years, ecological sciences have generated new and valuable tools for assessing and improving land, and new models of sustainable use of land have proved their worth.

After laying out goals and pledges, they concluded, "with the grace of good fortune, the West may finally create what Wallace Stegner called 'a society to match its scenery.'" Over time, more and more ranchers found that the Coalition's "radical center" offered them the home they had been seeking. Membership swelled, joint projects were embarked upon, books were published, and a high-energy annual conference was organized.

However, when they tried to expand to the larger ranches, the huge spreads that shuffle millions of cows into huge feedlots each year, they found zero interest. "It was like hitting a brick wall," Arturo Sandoval, board chair of Quivira, told us. "The corporations and big city businessmen who own the giant ranches just don't have the same feel for the land that the smaller guys do." Sandoval is a perfect example of this new mentality. A Chicano activist in his youth, Arturo was the western states coordinator of the first Earth Day in 1970. He has been a friend of Denis's for forty-five years. Now, having recently shorn his full beard and added a few pounds, he retains the twinkle in his eye and a robust sense of humor. Rolling his eyes as he describes himself as "a respected elder," Arturo devotes much of his time to organizing agricultural cooperatives. His youthful idealism is intact and his hope is undimmed, but he is now "much more interested in making progress than making noise." That, in a nutshell, is the Quivira Coalition mindset.

As water shortages and escalating grain prices change the economics of CAFOs, and as a robust market develops for lean, healthy grass-finished beef, many of the huge ranches will almost inevitably be converted into medium-sized ranches with diversified ownership. The Quivira Coalition will welcome them and eagerly share the lessons that it has learned.

Tom Steyer and Kat Taylor, the husband and wife who own the TomKat Ranch in Pescadero, California, were surprised by the high demand for their holistically raised beef. When they purchased the 1,800-acre property, they didn't have much interest in cows. Tom, the founder of a hugely successful hedge fund, and Kat, the president of a wildly innovative nonprofit commercial bank, planned to use the ranch to get hands-on experience with solar energy (Tom) and explore ways to restore the soil and conserve water (Kat).

Along the way, they discovered that Savory's intense rotational grazing methods would provide the water and soil benefits they were seeking while producing healthy beef. An affiliated chicken farm, Early Bird Ranch, operates mobile chicken and turkey coops that trail the cows from pasture to pasture, eating insects attracted by the dung, and then scattering their own dung.

Because the ranch is located in an area near Santa Cruz that is sometimes scorned by political conservatives as the "left coast," the two progressive capitalists branded their product LeftCoast Grass-Fed Beef. The cows are Black Baldies, an Angus-Hereford crossbreed known for producing tender beef on an all-grass diet. Demand for LeftCoast GrassFed Beef skyrocketed well beyond the ranch's ability to meet it at the limited scale the owners wanted. Tom, the hedge fund titan, pointed out that this is a classic indicator that that they were underpricing their product. Kat, with her nonprofit orientation, replied that they were making a more than adequate return on their investment and that she didn't want to grow healthy food just for the richest 1 percent.

Hardworking perfectionists who aren't easily intimidated, Steyer and Taylor are ideal proponents of the holistic approach. They set out to do everything right for the cows, the land, and the eventual consumers, and they hired people like themselves to manage day to day operations.

LeftCoast GrassFed Beef is 100 percent grass-fed and grass-finished. The cows are given no hormones or antibiotics. The beef is certified by the American Grassfed Association and the Food Alliance, and it is Animal Welfare Approved. Even the electric fences, needed for effective rotational grazing, are solar powered. The pastures use no herbicides or artificial fertilizers. The ranch is rich with abundant grasses and forbs (plants other than grass that lack woody stems) as well as diverse wildlife. And most of the product is sold within fifty miles. In a world in which the term "sustainable" has genuine content instead of just serving as a greenwashing tool, the TomKat Ranch will be the norm, not the exception.

Creative Ideas to Lessen Cows' Impact on Soil

A new field of research called "regenerative agriculture" focuses on how soil-building can reduce global warming and perhaps pull enough carbon out of the air and tie it up in soil to take us back to levels of atmospheric carbon dioxide prevalent in 1980.[55]

Plants and earthworms are key players in building soil. Plants draw energy from the sun through photosynthesis: They pull carbon dioxide from the air and tuck it into carbon-containing packets of carbohydrates. Some of the carbon goes into plants' roots, where friendly fungi and soil microbes latch onto it and store it in the form of humus. Humus provides channels in soil for air and water, and humic acid makes it easier for plants to pick up nutrients.

Properly managed grasslands can sequester substantial amounts of carbon, binding it up in soil and making a cow's pasture a "carbon sink."[56] The Plains CO_2 Reduction Partnership (a joint project of Ducks Unlimited, North Dakota State University, and the United States Geological Survey) found that pastures can be turned into carbon sinks by applying manure and introducing earthworms to soil, by seeding with a mixture of perennial grasses and alfalfa, and by limited, carefully timed grazing instead of mowing.[57] Leaving a residue of crop on a field also builds up carbon.

Another way to improve soil is to mix charcoal into it, if the charcoal is made in a sustainable way, for example from waste materials. Charcoal is basically wood and other vegetation that has been heated in a contained place without oxygen. This drives out moisture and chemicals that vaporize. When used to improve soil, charcoal is called "biochar" or "terra preta." Earthworms help mix it into soil. Biochar sequesters carbon for a very long time, and so might help with climate disruption. Properly employed, it can be a powerful soil builder. In the 1950s, the distinguished Dutch soil scientist Wim Sombroek discovered that indigenous peoples along the Amazon had used terra preta to build rich topsoil six feet deep in places where there had previously been only the very thin topsoil characteristic of the rain forest.[58] This is not only of historical interest. Sombroek's findings have stood the test of hard science and worked their way into modern agriculture and popular culture. Terra preta is now being used in many parts of the world to enhance impoverished soil.[59]

The Marin Carbon Project in California has experimented with applying compost to ranchland, compost that would otherwise have been dumped in landfills. Grasses grow faster on test plots and capture a metric ton of carbon per hectare (2.471 acres) year after year. Grazing cows cause grasses to grow new seed heads and not dry up as fast as they otherwise would.[60]

Plowing land and leaving it bare part of the year causes carbon to oxidize and float up into the air. In the Corn Belt, we've already lost nearly half of the carbon and nitrogen once in the soil.[61] One partial solution is no-till farming. No-till means planting crops in an unplowed pasture. Seed drills are used to insert seeds. Any needed fertilizer can be injected along with the seeds, rather than scatter-applied in pellets, thus reducing runoff. Residue from crops is left in the field.

When soil is treated this way, it retains more water, is less compacted, and is more likely to stay put. Nitrous oxide emissions are greatly reduced. The extra moisture in the soil might allow an extra crop to be grown during a year. The diversity of microbial, insect, and animal life increases. A considerable diversity of grassland species of birds have begun nesting in soybean fields that aren't tilled; the challenge is to find ways to time heavy machinery use to avoid crushing occupied nests.

A disadvantage is that weeds also grow better. Most farmers address this problem with herbicides, and they are strongly encouraged in this by Monsanto and DuPont. But there are more earth-friendly alternatives, such as cover crops. Done well, no-till can reduce or eliminate the use of chemical poisons.

In 2010 the USDA reported that over a third of the cropland planted with eight important crops was being managed as no-till.

Nearly half of soybean crops and a quarter of corn crops were grown this way, and the use of no-till was still increasing. The USDA believes no-till farming could significantly reduce our nation's contribution to climate warming.[62]

One of the most promising long-term prospects is greater use of perennial crop plants, plants that regenerate themselves year after year. Scientists think it's reasonable to aim for a life span in crop plants of three to ten years. An impressive array of agricultural scientists have concluded that better perennial grains are needed to help ensure food and ecosystem security.[63]

Perennials reduce erosion because their root systems have more time to grow and are therefore much deeper and more extensive than the roots of plants that live less than a year. Perennials' roots may penetrate twelve feet into soil, while annuals' typically reach only a foot or so. With perennials, land isn't left bare between crops. And because perennials don't have to be replanted every year, less fuel is burned for tractors and for shipping seeds. Clever mixes of perennial crops in the same field can reduce the need for the application of chemicals.

Planting perennials would save farmers money on fuel, seed, water, fertilizer, fungicides, and pesticides. A perennials farmer loses the ability to quickly change crops to time the market, but market timing doesn't seem to work any better for farmers than it does for IRA investors.

The catch? None of the big commodity crops are currently available in a perennial form.[64] It will take time and money to develop perennial crops because when plants put so much energy into developing root systems, they tend to produce smaller seeds. Thus, their yield is lower. Ed Buckler, director of the Buckler Lab for Maize

Genetics and Diversity at Cornell University, thinks this problem can be overcome by using genetic sequencing technology. Furthermore, in dry lands like much of Washington State's wheat country, the lack of water means annual crops can't be planted every year. So over the course of several years, perennials, with their deep, drought-defying roots, might produce more bushels.[65]

Much of the money for research and development in agriculture comes from corporations that sell seeds, pesticides, and machinery for growing annual crops. Annual crops are a form of planned obsolescence; corporations naturally favor research that will result in products they can patent and sell year after year.[66] Money for annual crop research is further facilitated by the USDA's (and Congress's) revolving doors to agribusiness.

The brilliant, indefatigable Wes Jackson—founder of The Land Institute in Salina, Kansas—has tried to stimulate serious academic research on perennials for decades. Although in most of his endeavors Wes is an irresistible force, his pleas within academia fell on deaf ears. So working privately and with modest funding, Wes and his colleagues have created new perennial crop plants by crossbreeding annuals with their wild relatives. They've been working on perennial legumes, mustard seed, sunflowers, rice, wheatgrass, and sorghum. Using selective breeding, they have made considerable progress with Kernza wheatgrass.

Other visionaries are tackling corn.[67] What we in the United States call corn is actually maize, which probably came from a small-seeded grain, teosinte. Teosinte's "corn cobs" are the size of Tootsie Rolls. Around 1100 BC, people living in Mesoamerica started tinkering with teosinte. Its kernels were tiny and too hard to chew, but if heated they popped like popcorn. Maybe this is what drew attention to the plant. It is thought that Mesoamericans saved seeds from plants that had larger ears and kernels, and maize eventually evolved and spread throughout the American continents.

Ed Buckler thinks it may be possible within twenty years to use new technologies to produce perennial corn with the same yield as today's annual crops.[68] His optimism (which is not universally shared) has an important caveat: There must be adequate federal funding. In 2009–2010 the USDA doled out a paltry $1.5 million in grants for perennial crop research,[69] and it doesn't provide the long-term grants necessary for such research. In a sane world, that figure would have been at least a hundred times higher. By way of comparison, from 2001 to 2014 the United States spent over $731 billion on Afghanistan.[70]

America has a breathtaking (to turn a phrase) supply of cow manure. Treating manure by composting it is an ancient practice. The most common method of making compost is to shovel manure into rows or piles and turn it over occasionally to aerate it. During composting, microorganisms in cow manure generate heat, which helps break down organic material and kill pathogens, fly larva, and weed seeds.[71] This reduces the need for fertilizers, herbicides, and pesticides. Composting also greatly reduces manure's volume, stabilizes it, and eliminates the stink.

Using compost instead of artificial fertilizer reduces warming gases. It also sequesters some carbon in soil. Compost is easier to handle than raw manure, and its application can be more even. Compost makes soil more suitable for planting and growing crops: It provides a good habitat for beneficial organisms, and makes soil less prone to erosion, more fertile, better aerated, and better able to retain water.

The barrier to large-scale composting is this: Most cows are now crowded together in giant confined-feeding operations, so the farmland around them can't absorb all the compost they produce. And it's too expensive to truck compost very far from where it is produced.

Smaller dairy farmers are experimenting with different types of composting.[72] One is Karen Bumann, of Wisconsin's Sweetland Farm, an organic operation. Driven, in part, by state regulations that now limit the amount of nitrogen that can be put on fields, she and other dairy farmers are trying a "composting bedded-pack" approach. When the weather's cold or inclement, cows are free to walk around a barn that has a large area sectioned off and deeply piled with bedding. Cows both lie on the bedding and use it as a toilet. The bedding might be sawdust, ground-up soybean, rye or wheat straw, sunflower hulls, or whatever's locally available.

Karen told us cows bedded this way seem happier, produce more milk, and live longer. Twice a day, the bedding needs to be stirred with a cultivator or chisel plow to mix in the manure and keep the surface dry. This constant mixing keeps it from being a true "compost" because it doesn't get hot enough for long enough. Some volatile materials still get into the air. The method does, however, generate enough heat to kill fly larva and organisms that cause mastitis.[73]

Dung beetles are the champs at converting cow pies into productive soil. They live for dung and can make cow pies vanish in less than a day. Because cows don't like to eat plants located close to their poop, the pasture area available for grazing quickly shrinks if pats aren't cleaned up. But the same mess that repels cows can attract dung beetles from ten miles away.

Dung beetles that specialize in cow poop so love the smell that they may hover near a cow's tail just hoping for a treat. When rewarded with a splat, they dive in while it's still warm. Although they prefer it fresh and stinky, dung beetles will burrow into hard dung that earthworms shun. They eat dung, squeeze water to drink from it, and bury their eggs in it to incubate.

Different types of dung beetles roll up balls of manure, tunnel through it, or dwell in it. The rollers, sometimes called tumblebugs, are the most visible. A male and female work cooperatively, using their back legs to roll a ball of manure. They bury the ball in soil and either eat it later or use it to incubate one of their eggs. After a month or so the beetles leave a patch and worms move in, filling tunnels with their castings.

If cow pies just sit on the surface and dry out, they lose most of their nitrogen to the air. By working dung into the soil, beetles improve its texture and recycle nitrogen and other nutrients needed to grow lush grass and forbs for cows to eat. Runoff is reduced because water and roots can penetrate better. As a toss-in benefit, the beetles kill the larva of certain flies that plague cows (and farmers), and they even reduce bovine gastrointestinal parasites by eating parasite eggs before they can turn into larva and infect other cows. It's estimated that dung beetles currently save United States cattlemen about $400 million a year in clean-up costs.[74]

Ancient Egyptians were deeply intrigued by scarabs, a type of dung beetle. They saw in scarabs a microcosm of the cycle of life, the daily rebirth of the sun, and the eternal nature of the human soul. You couldn't walk a block in ancient Egypt without encountering a dung beetle image in an amulet, adornment, seal, carving, or painting. Turns out dung beetles do indeed have a link to the heavens: researchers in Sweden and South Africa recently discovered that dung beetles navigate by reference to the Milky Way—the first animal found to do so.[75] During the day they navigate by the sun. Unlike humans, they can see and navigate by polarized light in the sky.

Dung beetles co-evolved with cows in Europe and Africa, and they come in eight thousand varieties. Hawaii and Australia were two places not blessed with them, so cattle ranchers had to import them. It worked beautifully; the imported beetles cleaned up the messes from the imported cows. Thanks to dung beetles, it's now once again legal, in parts of Australia, to dine at an outdoor café that's not protected by screens.

You can find a type of dung beetle that will tolerate just about everything, including heat, drought, and cold. Some beetles work at night, some during the day. Different species of dung beetles will work cooperatively in the same pasture. There's a growing body of scientific literature on how to match dung beetles to particular environments; getting the mix right is crucial for optimum results.

Unfortunately, the sheer number of cows crowded into factory farms creates terrible working conditions for the beetles. The pesticides, fungicides, and insecticides that saturate large cow operations are harmful to beetles. For example, although beetles can greatly reduce the number of bovine gastrointestinal parasites, beetles are killed by medicines fed cows to kill the parasites, because the active part of the medicine passes through cows intact. Residual antibiotics in cow poop kill off good bacteria on which dung beetles depend.

In short, it seems unlikely that even dung beetles can clean up the factory farm mess, although refining our use of them might greatly benefit smaller farms.

Worm Ranching

Like dung beetles, earthworms can clean up cow dung. Late in his life, Charles Darwin became downright obsessed with earthworms. He collected and weighed worm castings and reported, "[A]ll the vegetable mould over the whole country has passed many times through, and will again pass many times through, the intestinal canals of worms."[76] Worms break rocks into fine mineral soil, and worm castings slowly bury the ruins of lost civilizations. On some farms the earthworms under the soil actually outweigh the cows above them.[77] Tilling soil shrinks the number of earthworms (yet another argument in favor of developing perennial crops).

Could multitudes of worms be used to turn *mountains* of unwanted cow manure into fabulous soil enhancements? We dug into the topic. Like many critters, including humans, earthworms are basically a tunnel within a tube. Worms lack teeth but have mighty mouth muscles. Using the tiny bristles attached to most of the round segments that make up their bodies, they burrow their way into manure, mouths wide open. They prefer their food a little soft, a little rotten. As manure passes through a worm's inner tunnel it goes through a gizzard (like chickens have, a sort of grinding stone), and through intestines where nutrients are extracted. Microorganisms do much of the digestive work; worms themselves provide mostly mechanical grinding and mixing, exposing more surface area to microbial magic.

An Internet search revealed we weren't the first to wonder whether worms might turn dung into something more useful. We discovered there was a worm farm only a few hours' drive from our house. On a weekend in late August we headed south from Seattle toward the snow-topped, 14,410-foot-high Mt. Rainier. The community of Yelm is nestled just to Rainier's west. This country is stunningly beautiful. Yelm is thought by some to be home to Ramtha, whose followers believe is a thirty-five-thousand-year-old god channeled through the body of a sixty-something-year-old blonde woman, JZ Knight. Hmm. We were obviously in a part of the country that tolerated unusual beliefs. Would our worm farmer be a crackpot?

After driving past the spiritual compound, we came to a very down-to-earth place, the Yelm Earthworm & Castings Farm. Kelan Moynagh, its proprietor, greeted us outside his bare-bones office. An engineer by training, Kelan is a man of middle years, sunburned, with an affable demeanor. The worms churn out a profit for him, enabling a middle-class lifestyle.[78] His product, we learned, is in great demand by experienced gardeners and farmers.

"Worm ranching is animal husbandry. I learned what I needed to know from a little reading and high school biology," Kelan told us. To be blunt, earthworms turn cow poop into worm poop. This might not seem a heroic improvement, but it actually is. To make worm poop more attractive to shoppers, the product is referred to as "castings," "cast," or "vermicast." Kevin filled us in on castings' benefits. Unlike cow manure, castings don't stink. And castings are more beneficial to soil because they have a uniform composition rich in calcium, nitrogen, phosphorus, potassium, and humus. Worms quickly turn the nitrogen in manure into useful nitrates—without the need for high temperatures. Worms also reduce the amount of carbon dioxide released into the atmosphere.[79] Castings are said to be free of pathogens that sicken humans, although there's a split of opinion on this. There is evidence that vermicompost can protect plants from some plant pathogens, such as those that cause root rot, but the process is not understood.[80] Even after leaving the worm, composting activity continues for quite a while in the castings. Worm slime allegedly keeps nutrients from being quickly washed out, so vermicast is a "slow-release" fertilizer.

Researchers at Ohio State University report: "Vermicomposts can have dramatic effects upon the germination, growth, flowering, fruiting and yields of most crops, particularly fruit and vegetables, which are high value crops." They attribute this to plant-growth regulators that are produced by earthworms interacting with microorganisms.[81]

Kelan calls worm castings "gourmet compost," and claims that he has watched earthworms and their bacteria pals under a microscope kill *E. coli* just by touching it. A single gram of vermicompost (equal in weight to a small paperclip) contains a billion bacteria believed to belong to up to ten thousand different species! No one has the faintest idea how all these bacteria interact with soil, earthworms, and plants.

Red wigglers normally live in rotting vegetation near soil's surface. But they can thrive on fresh cow manure, which is free for the asking, and worm farmers can up their production by pampering them. Worms like their manure medium warm, not over 120 degrees Fahrenheit. Kelan showed us around six cavernous, dark rooms where the worms slave away. Long rows of castings about a yard high ran the length of each room. The wrigglers need only shelter, water, and manure, so Yelm Earthworm employs only two part-time human workers.

After about eight months, a row is ready to be harvested. The worm wranglers wet it down heavily, which drives the worms to the surface. After shoveling off the worms they can reach, the men load the rest into a separator, a rotating screen tube that filters out the castings, which look a bit like coffee grounds.

We moved on to a field and a greenhouse, where vegetables were being grown with castings. It had been a bad year for growers in Washington, but these plants were thriving. In addition to selling soil-enhancing vermicompost, Yelm Earthworm grows garlic for Whole Foods. We bought a bag of Kelan's compost, took it home, and used it to grow outstanding chard and flowers. Most of the organic farms near Kelan's worm ranch use Yelm Earthworm's product, and there are a few non-local bulk buyers.

Quality castings cost more than other fertilizer options, however, and transport costs are high. Another obstacle is that labeling laws differ from state to state and weren't written with organic products in mind, products that often vary somewhat from batch to batch in

The worms and this simple "separator" machine, which separates worms from vermicompost, do most of the work at Yelm Earthworm & Castings Farm. Photo: Gail Boyer Hayes.

their nutrients so a label can't reveal precisely how much nitrogen and other fertilizers the product contains.

But castings can be made into "vermicompost tea," an inexpensive mixture of castings, water, molasses, fish, and kelp, which can be sprayed on plant leaves or soil. The tea is loaded with beneficial bacteria, fungi, and protozoa, and wards off a long list of plant diseases. It can be sold in powdered form, which makes it less expensive to transport. Kelan told us that Harvard University uses his compost tea on its twenty-five acres of lawn, which has cut in half the need for irrigation. He said that the college intended to expand its use to all eighty acres of Harvard landscaping.[82]

The six hundred cows at nearby Plowman Dairy, which supply Yelm Earthworm with manure, create about sixty cubic yards of manure a day. Yelm Earthworm can handle only forty cubic yards every two weeks. Learning this led us to wonder whether vermicomposting could operate on a much larger scale.

The largest vermicomposter in the Western Hemisphere is Worm Power, in Avon, New York. Assisted by two grants from the USDA's Small Business Innovation Research Program, the company began operating in 2004 and is now churning out a profit for investors. Worm Power uses a patented, automated system that's higher-tech than Kelan's outfit.

Denis and the president of Worm Power, Tom Herlihy, had a long conversation over a Skype link. An opinionated, up-front guy, Tom is very well informed on the mechanics of vermiculture, holds a master's degree in agricultural and biological engineering from Penn State, and works with researchers from Cornell who are studying dairy waste. He jokes that his sixty million worms make him the largest employer in the country. When he lets his guard down, he refers to his castings as "sexy." Worm Power compost is USDA-certified organic and is consistent in its quality and content. It is a serious, hard-nosed business.

Waste is trucked in from a large nearby dairy farm. The liquid is squeezed out, goes into a lagoon, and then is put on cornfields. Unlike Kelan, Tom combines the solids with silage and composts the mixture, generating enough heat to pasteurize it and kill pathogens as well as many weed seeds. The finished compost is fed to the worms under carefully controlled indoor conditions.

This is Club Med for worms. According to Worm Power's website, they can turn 1,320,000 pounds of compost into vermicompost in about sixty days. The product is marketed across the country and trucked in bulk to large greenhouses, vineyards, and golf courses. Because the product appears to suppress many plant diseases, researchers at Cornell think it may prove an alternative to some herbicide and pesticide use.

Our conclusion: Vermiculture is a wonderful idea that produces a valuable product. However, it is not scalable to handle wastes at the amounts produced by big CAFOs. Still, we emerged vastly more impressed by vermiculture than we expected to be. Moreover, handling CAFO wastes is the wrong metric; we believe that CAFOs should go out of business. As one component of an integrated waste management system for sustainable dairy farms and beef ranches, worms may have an important role to play.

The Roundup

We uncovered some promising ideas that could lessen the pollution caused by ninety-three million cows and that wouldn't break the bank: adjusting cows' diets to reduce their emissions of warming gases, implementing better soil and water conservation practices on farms, making those who lease agricultural land share responsibility for protecting and improving soil, vermicomposting, and shifting to perennial crops and regenerative agriculture. Our sense is that it's too late to put the Ogallala to its highest and best uses (or even to agree on what uses those are), and that many farms and communities dependent on the aquifer will soon be hung out to dry.

The growing demand for foods that are organic, locally grown, and grown in sustainable ways should encourage more farmers to adopt good practices we uncovered. But even if implemented to their fullest practicable extent, these practices can only modestly reduce the environmental problems caused by keeping so very many cows.

A glimmer of hope does come from down under. Australia has twenty-seven million cows and sixty-nine million sheep. Aussie cattle and sheep producers have promised to be farming sustainably by 2020. Urban Aussies apparently don't like eating something that troubles their ethical, social, and environmental consciences.

The Aussie ranchers plan to protect more wetlands, waterways, and biodiversity; reduce emissions and the number of animals in a pasture; provide native vegetation corridors; and use soil more sustainably. A hundred research projects are under way. The president of the Cattle Council of Australia, Andrew Ogilvie, told the *Australian* reporter Sue Neales, "[C]onsumers rightly have expectations about how their food supply is produced, and it is up to livestock producers to demonstrate we are sustainable and moving even more in that direction backed by sound research."[83] Amen.

Chapter Three

HOOFPRINTS ON YOUR TABLECLOTH

Cows have an astonishing ability the brightest chemists haven't been able to replicate in the lab: They can convert low-value cellulose and carbohydrates (grass and by-products) into massive amounts of high-quality, delicious protein (milk and meat). Cows manufacture complete protein with all nine amino acids people require to thrive. To accomplish this bio-alchemy, a cow spends six to seven hours a day eating and another eight hours chewing cud—a fifteen-hour day, seven days a week. And a cow works for just room and board, with a bleak retirement plan.

This is the drill: Ferdinand wraps his three-pound tongue around vegetation and pulls it into his mouth. He swallows the greens without much chewing. A cow has just one stomach, with four compartments. When the first and largest compartment, the rumen, is full, Ferdinand burps "cud" back into his mouth. He grinds the cud between his powerful molars and the toothless but tough palate on the top of his mouth. (Gary Larson is the world's best cow cartoonist, but only cartoon cows have upper front teeth to flash when they smile. Real cows have a leather-like dental pad.)

The rumen is a universe unto itself, teeming with hundreds of species of bacteria, protozoa, and fungi. Its lining looks like tufted carpet. The late Lynn Margulis, a distinguished biologist, once described a cow as "a forty-gallon fermentation tank on four legs."[1]

The microbes cooperate and compete among themselves, using enzymes to ferment the mashed vegetation and to create protein, B and C vitamins, and volatile fatty acids. The rumen enables a cow to extract energy from plant matter like stems, seed coats, and seed shells that most other animals can't eat.

The second part of Ferdinand's stomach is his reticulum, which has a lining that resembles a honeycomb. Cows aren't picky eaters, so the reticulum acts like a trap. People poking around inside dead cows find rocks and bits of plastic and metal in a reticulum. "Hardware disease," as it's called, causes problems if metal punctures the stomach wall. To prevent this, a finger-sized rod ringed with magnets can be inserted into a cow's stomach where it attracts swallowed metal. The magnet stays in the cow all the way to the slaughterhouse. Of course, magnets won't attract glass, plastic, or aluminum. When people toss trash out of their car windows, the trash sometimes ends up in fields, gets shredded by harvesting machines, mixed in with hay and silage, and is eaten by cows. A moment of thoughtlessness can do a cow serious harm.

The third part of a cow stomach, the omasum, has folds that remind some people of the pages of a book. It filters and resorbs water. Finally, food gets to the part of the cow's stomach that most resembles a human stomach, the abomasum, where acid and enzymes digest protein. After the stomach comes the intestine. Unfurled, it could stretch to the height of a sixteen-story building.

Cows are amazing. They will always have a role to play on land that provides good pasture yet is unsuitable for growing crops. But cow are too high on the food chain to continue to play such a prominent role in American diets.

Eating Up the Food Chain

A food chain is a flow of energy from the eaten to the eater, usually from a little organism to a larger one. It starts with something green that harvests energy from the sun through photosynthesis. The green thing is often eaten by a herbivore, and the herbivore by a carnivore or omnivore. To survive, a predator must use less energy getting food than the food contains. Here's one simplified food chain: Corn > cow > you.

Because a great deal of energy is lost at each stage, food chains tend to be short. Some energy goes into making seeds and offspring. More energy is used for growth, movement, and reproduction, some for keeping an animal warm, and some remains in feces. This energy isn't really lost—it's still hanging around in some form as heat, chemical bonds, and such—but it isn't passed on up the food chain. *Typically, only about 10 percent of the energy at one stage moves on to the next.*

In November 2013, a leaked report by the Intergovernmental Panel on Climate Change said that climate change might cause a drop as big as 2 percent in global food production each decade. This is a grimmer prediction than earlier reports and is based on new research on how heat waves affect crops. At the same time, demand for food is expected to increase up to 14 percent per decade.[2] This means humans (and cows) will have to become ever more efficient in getting the calories they need out of food.

Today, most American cows are finished (fattened) on grain, not grass. It takes roughly ten thousand pounds of corn to produce one thousand pounds of cow, so grain-fed cows will become expensive luxuries in a crowded, hungry world. For humans, a vegetarian diet that includes beans, lentils, peas, and chickpeas can be just as nutritious as one based on beef, while feeding perhaps ten times as many people per acre of farmland.[3]

Grandma's and Grandpa's cows were benign beasts that transformed grass into protein, fertilized their own pastures, and provided agricultural muscle. Today's cows are more like energy-sucking black holes. How much energy they require depends on what you include in your calculations. But if you include all the fertilizer, pesticides, and fuel required by growing and harvesting cow feed, the total is daunting. David and Marcia Pimentel of Cornell University calculate that 40 calories of energy are used to produce 1 calorie of beef protein. The ratio for milk is 14 to 1.[4] And this calculation ignores the additional energy needed to process, refrigerate, transport, package, and cook the meat.

If you assume Ferdinand is *never* put in a feedlot, is given concentrated feed supplement (about 838 pounds' worth) only during winter months, weighs about 1,190 pounds when slaughtered, and that about 35 percent of him is edible, he needs only two pounds of concentrated feed to make one pound of boneless meat.[5] This is far better than a cow in a feedlot.

The higher you eat on a food chain, the greater your "carbon footprint." For purposes of this book, your carbon footprint is the sum of the greenhouse gas emissions that your diet contributes to the atmosphere. According to the United Nations Food and Agriculture Organization, the average American's yearly beef consumption produces as much greenhouse gas as driving an automobile 1,800 miles.[6]

There is no real dispute that, if you want to slow global warming, eating less red meat is critically important.[7] From a financial, nutritional, and environmental perspective, protein is the most expensive component of our diet. Gorging on beef, as Americans do, wastes money and energy.

Sweden is a leader in analyzing food footprints. Swedes calculated that a quarter of their per capita global warming emissions came from feeding humans. To reduce this figure, the country pioneered food labels for use in grocery stores and restaurants, labels that list the carbon dioxide emissions associated with each food. For example, the labels advise Swedes to favor carrots over tomatoes and cucumbers, because the latter must be grown in heated greenhouses in their cold country. But if the carrots are grown in peat soil, the ranking shifts, because so much carbon dioxide is put into the air when peat is plowed.[8] The Swedish government hopes that the labels will lead to a 20 to 50 percent reduction of emissions from food production.

Eating less-processed, in-season foods that are grown close to your home will also reduce your carbon footprint, but it won't make as much of a difference as eating lower on the food chain. Eighty percent of the carbon footprint of food goes into the air before food leaves the farm.[9] There are many important reasons, including freshness, to patronize small, local, organic farms. Eating locally reduces the energy used for transportation. But to reduce the climate impact of your diet, the most important shift is to reduce your red meat consumption.

Our pets—especially our dogs—also eat large quantities of beef. There has been little rigorous investigation of the carbon pawprints of commercial pet food. Robert and Brenda Vale, authors of *Time to Eat the Dog: The Real Guide to Sustainable Living*, calculate that an average-sized dog eats 3.17 ounces of meat a day. Over a year, that's about seventy-two pounds of meat—only marginally less than the average person, globally.[10] A cooperative effort between the Nutro Company and the University of Illinois affirmed that dogs and cats require particular nutrients, not particular ingredients, so that plant protein, and protein from lower-order animals or even single-cell organisms, can be substituted for red meat in cats' and dogs' diets.[11] Furthermore, an estimated 57 percent of American cats and 53 percent of dogs are overweight. Twenty-seven percent of cats and 17 percent of dogs aren't just overweight but are obese[12]—as are nearly 36 percent of their owners.[13] The family can all go on a diet together.

How Corn Conned Cows

By the time Pilgrim feet tapped Plymouth Rock, corn (maize) was already the main crop cultivated in the Americas. Many varieties of corn had long been grown from Chile to North Dakota.[14] The Pilgrims wouldn't have survived without corn. For cows, on the other hand, it was the beginning of a long period of misery.

Humans probably started farming corn in the Western hemisphere before their distant cousins on the other side of the globe began domesticating cows. Just as cows became the most thoroughly domesticated animal, corn became the most thoroughly domesticated crop. Both now depend on humans to perpetuate their existence. (Without human help, corn kernels don't just fall off their cobs and plant themselves.) But for eons, cows and corn had no contact with each other, no need to evolve to accommodate the other's needs.

Unlike other major grains, which were exported to the New World from Europe, on either his first or second voyage Christopher Columbus came upon corn in Cuba and brought this novelty back to Europe. The cultivation of corn quickly spread throughout Europe and into China. The result was the collision of what may be the two most important human-sculpted organisms: corn and cows.

The secret to corn's success is that it is an energy-stuffed C4 plant.

Such plants are able to pack four carbon atoms into a compound during the first stage of photosynthesis, while C3 plants (the vast majority of plants) can take in only three carbon atoms. C4s don't have to open their stomata (air pores) as much to suck in carbon dioxide, which gives them a stem up in hot, dry climates.[15]

You'd never know it, driving through cornfield after cornfield of identical plants, but there are more than 100,000 pedigreed types of maize.[16] Corn has 32,000 genes, more genes than people have (around 20,000), and all those genes make the plant amazingly adaptable.[17] Exposing corn to radioactivity during atomic bomb tests in the 1940s, for example, created many weird varieties. Two universities in Iowa maintain seeds from most corn varieties. The mutants are useful mainly for basic research, but some have been, or will be, commercialized.[18] This is good—corn will have to adapt to a lot of climate disruption in the years ahead.

Lured by lucrative federal subsidies, farmers in the United States planted 97.4 million acres of corn in 2013.[19] These monoculture crops weren't headed for your dinner plate; the bulk of the crop is used to fatten cows or to make ethanol.[20] A popular crop is a type of field corn called "number 2 yellow dent corn," because each kernel has a little dent on both sides. Field corn can chip a human tooth—it's that tough. To consume it, humans must grind it into meal or treat it with lye to remove the thick outer skin. The sweet corn you buy as corn on the cob has a very thin skin around each kernel.

Although most of the field corn crop goes to cows and cars, you're downing a lot more field corn than you might think, often in the form of high-fructose corn syrup. Table sugar (sucrose) is more expensive in the United States than in other countries because of tariffs and quotas that protect domestic sugar producers, and corn syrup is much less expensive because of rich federal subsidies for corn. So the cheapest way for the American food industry to appeal to our collective sweet tooth is with high-fructose corn syrup. Four thousand processed foods found on grocery store shelves contain corn. Corn can even be found in instant coffee, aspirin, and potato chips.[21]

As anyone knows who has watched dairy cows lick up the grain they are offered as an inducement to behave nicely while being milked, grain is like candy to cows. Cows gain weight faster when fed field corn than when pastured, and their flesh has more marbling.

But cows on a corn diet suffer constant indigestion. Cows need roughage; a lack of fiber for fiber-degrading microorganisms to feast on causes acids to accumulate in the rumen, turning its contents from neutral to acidic. Slime forms and coats the rumen. Gases can't get out so they build up and constrict a cow's heart and lungs. (It takes careful management of cows' diets in a feedlot to reduce this danger.) Excessive acid also provides a good environment for unwanted bacteria to grow and cause ulcers.[22] Other bacteria can then pass through the ulcers and get into cows' bloodstreams. From there, germs travel to cows' livers, where they make pus-filled holes called abscesses.[23] In a PBS *Frontline* interview, the CEO of a huge cattle-feeding business in Kansas City said that because beef livers don't sell for much in the United States, and because corn makes the overall animal grow so much faster,[24] it doesn't matter if liver quality is diminished, or if liver has to be discarded.

When cows are fed grass they seldom need antibiotics, and there's little risk of bad types of *E. coli*, such as *E. coli* O157:H7, being passed along to humans. But unwanted bacteria flourish in cows on the cob, so the cows are given constant low doses of antibiotics.[25] Traces of these drugs end up in meat, milk, and the environment. Using antibiotics in this way also encourages the development of antibiotic-resistant bacteria.

• • •

Until the twentieth century, cows seldom overdosed on corn. The finger of blame might point to Fritz Haber, a German scientist, but cows were the last thing on Haber's mind as he toiled in his lab. Early in the twentieth century, he was awarded the Nobel Prize in Chemistry for inventing a method to make ammonium nitrate from the abundant nitrogen in the atmosphere (the air we breathe is about 78 percent nitrogen). Haber's colleague, Carl Bosch, found a way to dramatically scale up Haber's tabletop experiment into a full-blown industrial manufacturing process and make it economical. Much later, Bosch, too, won the Nobel Prize in Chemistry, and he went on to found the huge chemical company IG Farben. The Haber-Bosch process now produces one hundred million tons of nitrogen fertilizer every year. Ammonium nitrate fertilizer revolutionized agriculture, winning a temporary respite in Malthus's race between human population growth and food limits. The world's human population today might be a third smaller but for the extra food made possible by this fertilizer.[26]

Ammonium nitrate not only caused a burst in crop growth; the chemical is also used as an oxidizing agent in explosives. Unlike the British, who owned natural deposits of the ammonium nitrate in Chile, the Germans had no good source. World War I could not have achieved its scale had Haber not found a way to synthesize the chemical.[27] For a while, therefore, ammonium nitrate was killing more people than it was feeding. Even today, it is a critical component in improvised explosive devices (IEDs) and ANFO—a mixture of ammonium nitrate with fuel oil favored by terrorists at home and abroad.

In the United States, ammonium nitrate is entangled with the history of a huge munitions factory built on the Tennessee River in

Muscle Shoals, Alabama. The factory was intended to make explosives for use in World War I, but the armistice was signed two weeks before the first explosives were completed. World War II provided an opportunity for the plant to fulfill its original mission. After that war ended, the factory and electricity-generating dams were turned over to a federally owned corporation, the Tennessee Valley Authority, and the factory converted to churning out fertilizer.

The crop most suited to utilize this fertilizer was nitrogen-hungry hybrid corn. Even before the world wars, in the early 1900s, experiments were under way in the United States to further improve maize as a crop. A maize plant has both male and female flowers. The tassel on top of the plant is male; the silks on each cob are female. If maize self-pollinates, each generation of plants becomes weaker and has smaller cobs. But if two genetically distinct lines of corn are bred, for reasons still not fully understood "hybrid vigor" (heterosis) increases yield.

Hybrid corn has stalks that are less likely to break below the ear, and hybrid plants conveniently all mature at about the same time, making it easier to use large harvesting machines. In the 1940s and 1950s, yields of grains began to soar, from a bit over 20 bushels per acre in the early 1900s to over 155 bushels in 2013.[28] Wow. Hybrid seed is patentable. If a farmer tries to grow a new crop from the seeds of the previous year's hybrid plants, she not only will see a big reduction in yield but also will be breaking her contract with the seed company. So she needs to buy seed year after year.

Other downsides to hybrid corn are that it requires huge amounts of fertilizer, and the loss of biodiversity can increase vulnerability to corn diseases, such as happened during the outbreak of Corn Leaf Blight Race T, a fungal disease, in 1970.[29]

Corn wasn't the only crop being tinkered with in the 1940s. A young American researcher who had a PhD in plant pathology and genetics from the University of Minnesota, Norman Borlaug, went to Mexico to do agricultural research. While there, he developed new varieties of high-yielding, fertilizer-loving, big-seeded, disease-resisting dwarf wheat plants that (along with irrigation) greatly increased the amount of food that could be grown on a given amount of land.

These new corn and wheat seeds—and the attendant use of fertilizers, pesticides, farm machinery, irrigation, and larger-scale farms—were spread around the world by the Rockefeller Foundation. The Ford Foundation pitched in later, helping to spread high-yielding rice and corn grains. The burst in food production came to be called the Green Revolution. Food shortages had been a persistent source of international tension, and Norman Borlaug became the only agronomist ever to win the Nobel Peace Prize.

The Green Revolution resulted in a cornucopia of corn. The abundant corn, in turn, made possible the rise of giant confined animal feeding operations.

In the nineteenth century, when the practice developed of moving cattle to railroad holding pens in Abilene, Kansas City, Dodge City, and elsewhere, cows faced a new challenge. Traveling by rail was hard on them. Even when they didn't get sick or die en route, they inevitably lost much weight. So to prepare them for railroad shipment to distant slaughterhouses, the cows were penned for a few weeks and fattened on grain. By 1870, three hundred thousand head a year were passing through Abilene alone.

In 1876, Gustavus Swift, a former farmer and butcher, pulled together the elements (feedlots, slaughterhouses, railroads, and refrigerated railcars) that transformed the beef industry. His key idea was to slaughter cows in the west (Chicago) and ship refrigerated meat instead of live cows (only two-thirds of a cow being edible). This allowed Swift to fatten the cattle for slaughter rather than fattening them to make the trip.

Swift's approach, combined with the "disassembly-line" industrialization of the slaughterhouse that he pioneered, led to spectacular financial success, and Swift established the template for today's feedlot-dominated beef business. As fertilizer became cheap, as ultra-starchy number 2 yellow dent corn replaced the sweet corn Swift had used, and as antibiotics made it possible to control disease in huge, confined operations, the factory farm was conceived.

The final transformation took place under Earl Butz, Secretary of Agriculture in the Nixon administration. Butz was raised on a 160-acre farm in Indiana that was mostly self-sufficient. His family used a team of horses, not a tractor, and manure from livestock, not synthetic fertilizer. They grew all the animal feed they needed. After Butz left to attend Purdue University, a neighbor bought the Butz farm, and it grew in size to eight hundred acres. He purchased seed, fertilizer, pesticide, tractors, and feed—and yet made enough money to buy new cars, televisions, and vacations. The financial success of this approach to farming deeply influenced Butz's vision of American agriculture.[30]

In 1972, the Soviet Union faced one of its perennial grain shortages. It made large purchases of American wheat and corn while prices were still low, working through multiple exporters to hide the total volume of its purchases. When news of the purchases got out, American grain prices skyrocketed, and with them, the price of beef. The consumer backlash caught Butz by surprise, and he feared political repercussions. This led to the 1973 Farm Bill, the final element underpinning the modern industrial cattle business.

Under this legislation, the government set a price for grain and guaranteed the purchase of anything the market didn't buy. By removing most of the risk from planting, the legislation essentially guaranteed huge surpluses. To soak up the surplus grain, the USDA encouraged gigantic feedlots. Today corn comprises over 95 percent of all the feed grains produced in America, the others being sorghum, barley, and oats.[31]

Holsteins became the most popular dairy cows partly because of the breed's ability to tolerate corn. (Holstein's appearance might also be a factor in their popularity. Their cheery black-and-white spots—as unique to each cow as fingerprints are to humans—are purely the result of breeders' aesthetic preferences.) Holsteins on a feedlot can stay alive on an all-corn diet.[32] Although cows are stars when it comes to the *grass*-conversion business, compared to other animals they are clunkers at turning *grain* into meat.

The United States produces more corn and more beef than any other nation. The two facts are obviously connected. Today, because corn is also used to make ethanol, the price of fuel is closely linked to the price of corn and beef. That makes the United States even more vulnerable should our monoculture corn crops fail.

Corn ethanol began as an opportunistic initiative by Archer Daniels Midland (an Illinois-based global food processing and commodities trading corporation) during the Carter administration. The firm sought to capitalize on the emotions stirred by oil shortages relating to the Iranian Revolution. The initiative was not intended to become a significant program. Rather, it was supposed to offer an opportunity to gain experience with ethanol fuels and learn if they caused problems for cars, gas stations, and so forth, while cellulosic ethanol technologies were developed. (Cellulosic fuels are made from the inedible structural material in plants.) Some players on the national energy chessboard also viewed it as a tactic to ensure farm-state support for research on other biofuels. Thirty years later, cellulosic ethanol has yet to make a commercial appearance, and corn ethanol receives billions of taxpayer dollars annually for a product that, almost everyone agrees, makes no sense.

Over the last few years, more and more of the corn crop has been used to make ethanol. The shift to ethanol isn't due to the invisible hand of market forces but rather to the fact that our federal government heavily subsidizes ethanol and mandates its use in gasoline. Recently we reached a tipping point: late in 2012, more corn (40 percent) went into making ethanol than into domestic animal feed.[33] In November 2013, the EPA finally proposed reducing the amount of ethanol that must be mixed into gasoline, a proposal that was immediately denounced by the Renewable Fuels Association.[34]

After thirty years of consistent, generous federal subsidies, there remains a vigorous debate as to whether ethanol from corn even contains as much energy as is required to grow and transport the corn, operate the distillery, and so on. Professor David Pimentel of Cornell University is probably the most outspoken skeptic. In 1979, he was asked to chair an advisory committee for the Department of Energy to study corn ethanol as an alternative to gasoline. His committee concluded that corn ethanol was a net energy loser. Pimentel has stayed on the case for three decades, now claiming that ethanol production from corn uses 29 percent more energy than the fuel contains.[35] Michael Wang at Argonne National Laboratory and David Morris at the Institute for Local Self-Reliance independently make a convincing case that ethanol from corn is not a net energy loser and that Pimentel has not factored in recent improvements in efficiency. Pimentel doesn't budge an inch and retorts that his critics are ignoring the energy costs of labor, irrigation, and the manufacture of farm machinery.[36] Life-cycle energy analysis is a complicated business, and the arbitrary setting of boundaries can determine the outcome.[37] Even people who completely agree on every number can strongly disagree about the conclusion.

At some level, the whole controversy is ridiculous. Even if corn ethanol *does* produce a small net energy yield, it is far too trivial to make a meaningful contribution to our national energy needs. Wang and Morris may disagree with Pimentel's analysis, but neither of them is a full-throated advocate of corn ethanol. The super-efficient production of corn ethanol might be of some value to small farms and rural coops that use it close to where it is made, but at a national level it is unwise.

Growing corn for ethanol has had unanticipated side effects. Because America also imports ethanol to meet its mandated quotas, it has led foreign farmers to clear forests to grow corn. An impressive study concludes that, because of this, over a thirty-year period corn-based ethanol will double greenhouse gas emissions and increase greenhouse gases for 167 years.[38] Other scientists predict nitrogen pollution of the Gulf of Mexico will jump by a third by 2022 if ethanol production from corn continues.[39] Even though hybrid plants gorge on ammonium nitrate fertilizer, corn absorbs less than half of the nitrogen that's applied.[40] The rest runs off and gets into rivers, lakes, water tables, and the Gulf of Mexico. In the gulf, it helps create a nine-thousand-square-mile dead zone each year that kills fish, shrimp, and jobs.

It May Be a Choice Between Fewer Cows or Chaos

Photosynthesis, the method by which plants harvest the sun's energy and turn it into carbohydrates, makes plant and animal life possible.[41] Life is fueled in large part by carbohydrates. Humans and our herds of domesticated animals have driven demand for carbohydrates to unprecedented levels. Worldwide, in 1900 there were 450 million head of cattle and water buffalo; in 2000 there were 1.65 billion bovines.[42] Beef production is undermining the entire natural world—squeezing out wild animals, shredding ecosystems, slashing biodiversity. Species are now disappearing at the rate of one species lost every twenty minutes, a faster rate than at any time since the Cretaceous extinction event sixty-six million years ago.[43]

The Global Footprint Network calculates that "it now takes the Earth one year and six months to regenerate what we use in a year."[44] In other words, people and their bovine accomplices are eating up Earth's capital at an alarming rate.

Canadian Vaclav Smil is a prolific writer and one of our favorite curmudgeons. Until recently he was a Distinguished Professor at the University of Manitoba. Smil digs deep and does his near-obsessive research with a precision many researchers lack. We often disagree with his opinions, but his facts are reliable. Some of his statements are shockers: "The biomass of domesticated land animals, dominated by cattle, is now at least twenty times larger than the zoomass of all wild vertebrates." As Smil sees it, "If extraterrestrial visitors could get an instant census of mammalian biomass on the Earth in order to judge the importance of organisms simply by their abundance, they would conclude that life on the third solar planet is dominated by cattle."[45]

"Primary production"—all the solar energy captured by plants and algae—is the bottom link in every food chain. Smil has calculated that humans and our farm animals consume 40 percent of Earth's net primary production. Only 60 percent remains to support all other life on the planet—not just the vertebrates but also the beetles, ants, worms, octopi, jellies, clams, sponges, lobsters, spiders, and so on. If the human population grows another 50 percent, and everyone starts to eat as much beef as Americans, we will wipe out much of the rest of the animal world.

In 2008 and 2011, global food prices spiked to record levels because of weather disruption, diversion of food crops to biofuels, and market jitters.[46] From 2008 to 2014, Americans, comprising only 4.5 percent of the world's population, grew 33 to 56 percent (depending on the year) of all corn, along with a disproportionate share of all wheat.[47] Our Great Plains is now breadbasket to the world.[48] For much of the first decade of the new millennium, the demand for corn, wheat, rice, and soybeans in many parts of the world greatly exceeded the available supply. As a result, food riots and political destabilization occurred in nations as disconnected as Mexico, Uzbekistan, and Yemen. Food shortages helped trigger the Arab Spring. Researchers at the New England Complex Systems Institute have correlated the dates of riots with rising prices of food. "Social disorder is contagious," says physicist and systems scientist Yaneer Bar-Yam. "The more we see it happening elsewhere the more it becomes imaginable where one lives."[49]

Today, 842 million people go to bed every night hungry, far more than before the Green Revolution.[50] The Food and Agriculture Organization says that in spite of the growing world population there are still enough calories in the food being produced to feed everyone, but poverty and the failure (or inability) of developing countries to invest in land management and agricultural infrastructure deny people access to food. The June 2013 United Nations report, "World Population Prospects," predicts the global population will grow from 7.2 billion today to 9.6 billion by 2050 and 11 billion by 2100. Even this midrange projection assumes stunning improvements in the availability and acceptability of contraceptives and new opportunities for women outside the home in developing countries, especially in Africa.[51]

In the United States, people seldom starve to death. But many are "food insecure." In 2012, over 5 percent of Americans visited a soup kitchen or food bank at least once. The Supplemental Nutrition Assistance Program (SNAP, formerly called food stamps) provided food to 44.7 million low-income Americans in 2011.

For health, societal, and national security reasons, we believe it's important for our country to continue to produce enough food to feed all Americans at affordable prices. Because cows require such a large portion of our resources, keeping too many cows could make it impossible to do this.

At one time we were a nation of small farmers. Today, most Americans don't have access to enough land to grow their own food, so they depend on distant farms to supply their supermarket. We recently read an old book, *Pleasant Valley*, about the restoration of a farm in Ohio. In this book, Louis Bromfield, a widely popular writer in the 1930s and 1940s (the time of the Great Depression, the Dust Bowl, and World War II), argues with passion that good soil and family farming are the most basic needs of a wealthy, healthy nation. He makes a compelling argument that an individual is only truly secure when he or she is able to grow enough food to feed a family in bad times. Seventy years ago, Bromfield and others worried about how much American soil we'd already lost and argued for techniques similar to the no-till farming now being rediscovered by the USDA.

There's no way to empty our cities back into the countryside, but we can greatly reduce the possibility of future food scarcity by cutting back on our consumption of grain-fed cows. This will encourage the development of more diverse farming and the production of more food for people from the same amount of land. Instead of monoculture dent corn, more fruits, vegetables, and grains for human consumption can be grown.

As a nation, we eat more cow per capita than do the citizens of any other country except Argentina and Luxembourg.[52] By dining so heavily on beef, Americans are not only eating high on the food chain but are eating up the food chain itself. We are consuming the environment—the dirt, water, and temperate weather—that keeps us alive. That makes about as much sense as gnawing on our knees for nourishment.

Chapter Four

LAGOON BLUES

Denis lifted the cow waste to his nose, sniffed, and smiled. He crumpled it in his hand, let it run through his fingers, and admired its dusty, pleasing texture. This cow splatter had been treated by a process new to American farms, and still quite rare. We'll take you to the farm at the end of this chapter.

Cow Factories

Across America tens of thousands of ponds called lagoons[1] dot the rural landscape. The word "lagoon" conjures up images of cerulean blue ponds ringed by reefs and palms. But the lagoons in rural America are nothing like the Blue Lagoon in which Brooke Shields frolicked. They are where farmers store animal sewage. Brimming with chemicals and disease organisms, these lagoons fester like open sores. They are typically man-made ponds rimmed by dikes of scraped-up dirt. Lagoons are seldom covered, but there are many variations: covered or uncovered, lined or unlined.

Livestock's untreated sewage is America's least-regulated source of pollution. According to USDA and EPA estimates, concentrated livestock and poultry operations produce three times as much raw waste as do humans.[2] And each ton of raw manure is up to 110 times more polluting (in terms of biochemical oxygen demand) than raw municipal sewage.[3]

The biggest change in bovines' lives since the invention of the milk pail began when, in the mid-twentieth century, cows began to be moved off fields and packed in huge numbers into relatively small spaces. More than half of our animal-origin food now comes from concentrated animal feeding operations, or CAFOs (pronounced KAY-foes), and the percentage is growing.[4]

On smaller farms with land enough to absorb their cows' waste, lagoons can be managed so they aren't a problem. (It takes only one or two cows to provide ample manure to fertilize an acre, an area about the size of a football field.) But on CAFO factory farms that have far more cows than acres, it's often necessary to build multiple massive lagoons.

Whether you call them cow pies (American English), cowpats (British English), cow chips (if dry), meadow muffins (from bison), or nik-nik (modern Sioux for the feces of any bovine), cow feces pose an awesome challenge to the dairy or feedlot owner. Out in a pasture, cow pies just add punctuation to the scenery and fertilize the forage. But when we stood inside a barn containing many hundreds of cows, whenever a cow let loose it sounded like an old high-tank toilet flushing. It went on and on. Behind the cow a pile formed of what looked like an entire family-sized portion of brownie batter.

Cows are large, hungry animals. A lactating dairy cow needs about one hundred pounds of feed a day. Beef cattle are no slackers either. They produce more urine and feces, per pound of weight, than any other meat animals.[5] After cows extract from their feed what they need to grow, walk around, produce milk, and hang out with other cows, there's still a lot left over. Therefore, cows poop a lot. Maybe a dozen times a day. They go so often that there are popular contests called cow-pat bingo. Chalk lines divide a yard into maybe a thousand numbered squares, and a cow or two is put into the yard. People bet on which square will get the first splat.

If CAFO farmers just let manure pile up in barns and feedlots, they'd soon need ladders to get to their cows. So cow waste is scraped or washed out of dairy barns as slurry and piped into lagoons. From the lagoon, waste may go into a spray-field system and be splattered over nearby acreage. Or liquid is allowed to evaporate and every once in a while matter too solid to be sprayed is scraped out and put on fields, often resulting in a deposit of excessive nutrients (nitrogen, phosphorus, or potassium). The excess nutrients run off and pollute.

But farmers have been shoveling cow muck ever since Eridu, Uruk, and Ur, so how dangerous can it be? Yesterday's cow poop was different. When you cram tens of thousands of cows together, as some CAFOs do, germs thrive. So antibiotics are given prophylactically and constantly. To maximize profits, the cows in a feedlot must be fed rich grain and are often given hormones. Feed grain must be grown elsewhere, nearly always with heavy applications of artificial fertilizers and pesticides, and then transported to the CAFO. All this fertilizing and transporting requires tons of fossil fuel. Vast amounts of water must also be pumped in for drinking and cleaning up. Polluted water then has to go somewhere.

High-volume, profits-over-quality operations turn cows into fungible widgets, cogs on a production line. "Factory farming" isn't just a metaphor. Although Henry Ford is often credited with inventing the assembly line, he says in his autobiography, *My Life and Work*, that he borrowed the idea of moving conveyor belts from the livestock "disassembly" lines set up in Chicago by meatpackers. In factory farms, everything is streamlined and sped up: standardized feed, mechanized feeding and watering, and slaughtering. High-skill, independent farmers have been replaced by low-skill, low-wage employees.

CAFOS are often initially welcomed by states as a source of anticipated income. When California imposed more stringent regulations on CAFOs, Idaho opened its arms wide to beef and dairy feedlots. Law after law was changed in Idaho to make polluters feel welcome.[6] Not only were the Idaho nutrient management requirements very modest, but they were (and are) secret! Under Idaho state law, "the nutrient management plan, and all information generated by the beef cattle feeding operation as a result of such plan, shall be deemed to be trade secrets, production records or other proprietary information, shall be kept confidential and shall be exempt from disclosure."[7]

Neighbors of the new arrivals organized to fight the stench and to protect the purity of their drinking water and the value of their own real property. They complained that it wasn't right to have to look at thousands of cows in dusty, filthy lots without a single leaf or blade of grass. The protesters soon learned that concentrated cows meant concentrated wealth, with the political power that accretes when corporations are "persons" under law and can therefore buy votes with their money.[8]

A wealthy Idahoan was already at the head of the parade when new dairies arrived. J. R. Simplot, the world's largest supplier of frozen French fry potatoes to McDonald's, had decided decades earlier to make a run at the rest of the Happy Meal. He realized cows could be fed much of the waste produced by plants processing his potatoes. A true self-made man, Simplot left school and home at fourteen, in 1923; by the time of his death, in 2008, he was a billionaire and the most powerful man in Idaho. His attitude was that of the cowboy:

America has so much open space that nature can clean up after cows. As he grew his fortune and supplied the public with the cheap beef it demanded, Simplot often left it to others to live with, or clean up, the waste his cows created.

In an admiring 1998 profile of Simplot in *Range* magazine, editor C. J. Hadley compared him to the John Wayne character in the cattle-drive movie *Red River.* Speaking of environmental regulators, Simplot told Hadley, "They are putting the heat on me. I was dumpin' in the river for years and years and I got by in good shape but now I've spent millions cleanin' the water up. . . . Mighta killed a few fish and suckers but never hurt anything. They're blowin' holes around the feedlot in Washington [state] to see if I'm pollutin'. . . . Maybe we won't be able to feed cattle anywhere any more, I don't know. The problem is there's too many goddamn regulations." The J. R. Simplot Company's two enormous feedlots fatten nearly a quarter-million head at a time. Two hundred fifty thousand cows generate an awesome amount of waste, and 130,000 of the cows were in the Grand View, Idaho, feedlot, which did not treat its wastewater. Simplot was releasing 1,500 gallons of untreated wastewater every minute, 24/7, into the Snake River. In 2010, after years of wrangling, the EPA ordered Simplot to stop, and Simplot complied.[9]

When it comes to factory farms, Texas law is weak and enforcement is lax. Texas produces twice as much feedlot waste as any other state. According to the Southwest Regional Office of Consumers Union, waste pollutes 388 miles of Texan streams and 23,000 acres of Texan lakes.[10]

In Idaho, California, Texas, and elsewhere, beef CAFOs are located far from cities, so most Americans don't see or smell them and therefore forget about them when they bite into a steak or cast a ballot. CAFO dairies are often located somewhat closer to cities, because of milk transportation costs. Only a quarter of large dairies have enough land on which to spread manure and still comply with the 2003 Clean Water Act's nutrient application standards for nitrogen. If a strict phosphorus-based standard is used, only 2 percent of large dairies are in compliance.[11]

Early in the twenty-first century, the EPA began requiring CAFOs that discharge waste into rivers, streams, or lakes to apply for a National Pollutant Discharge Elimination System (NPDES) permit.[12] The NPDES regulates "point sources" of pollution, such as pipes or man-made ditches that funnel discharges into surface waters. It also regulates animal waste and wastewater that is spread on land for "nonagricultural" reasons (just to get rid of it) and to spills from lagoons. But neither the EPA nor any other federal agency has gathered the necessary data to enforce this law.[13] The Natural Resources Defense Council (NRDC), the Sierra Club, and the Waterkeeper Alliance sued to force the EPA off its duff, and won a settlement under which the EPA was ordered to track down factory farms and either verify that they are exempt from regulation or start regulating them.[14] But on July 20, 2012, the EPA quietly put a note in the *Federal Register* claiming the settlement was only a proposal, not a final agreement, and that they weren't going to require CAFO information to be submitted pursuant to a rule.

To get an NPDES permit, a farmer must comply with effluent limits and standards. Therefore, most big outfits should be packing up a lot of their manure and hauling it to fields, pastures, forests, or highway margins where it won't exceed nitrogen and phosphorus limits. But raw manure is expensive to truck around, and farmers can't economically haul it more than about six miles. Unfortunately, neighboring farms often have their own cows or prefer the convenience of commercial fertilizer, which doesn't stink, contains consistent amounts of nutrients, and is easier to apply (although it costs more and doesn't improve soil quality as does manure).

Even accidental discharges are illegal under the NPDES. The penalty is a fine. The EPA has the right to prosecute and occasionally does.[15] More often, the power is delegated to state agencies, which are unable to overcome the clout of the farm industry in state legislatures. An example is found in Iowa, a state with heavy runoff from factory farms. Iowa Citizens for Community Involvement, the Sierra Club, and the Environmental Integrity Project petitioned the federal EPA to take over enforcement of water laws from the Iowa Department of Natural Resources (DNR). The EPA agreed that the farms needed permits, and it ordered the Iowa DNR to inspect thousands of farms over five years or hand over power to the EPA. It is a victory, of sorts, but few expect the Iowa DNR to find religion in the next few years, and at the end of the five years, it is uncertain that Congress will appropriate the money the EPA needs to do the job.[16]

Pollutants in Modern Manure

Because we have stuffed staggering numbers of cows into factory farms, we Americans now live in a cow toilet. Pollutants in feedlots and lagoons rise into the air and travel long distances on the wind. They also sink into groundwater, where they go with the flow. Managing a feedlot is a balancing act between not keeping the land too dry or too wet. If it's too dry, the wind kicks up noxious dust. If too wet, the stench and fly population balloon.

When we drove south on Interstate 5 a few years back and passed through central California, we knew we were closing in on the Harris Ranch miles before we could see it. As the ranch's website says: "Harris Farms . . . is California's largest cattle feeder, fed beef processor and beef marketer."[17] Our appetites went out the window as a stench like that of rotten eggs and ammonia poured in. This part of Interstate 5 is one of the few places where every day tens of thousands of Americans come nose-to-stink with the conditions that produce our cheap beef. The lagoons are on the far side of the cows, too far back to see from Interstate 5, but they stand out clearly on Google Earth.

Ammonia and hydrogen sulfide go into the air at every stage of manure handling: while manure is on the ground or the barn floor, from lagoons, when manure is scraped into a pile and stored, while it's applied to land, and while on a field. Forty to 60 percent of the nitrogen in cow food ends up in the air in the form of ammonia.[18] A single cow's daily burps include ammonia, hydrogen sulfide, volatile organic compounds, and as much as a thousand liters of methane, carbon dioxide, and other warming gases. Multiply that times 93 million cows!

Ammonia irritates the lungs of people and cows and makes it difficult to breathe. Over a quarter of factory-farm workers have respiratory problems.[19] Volatilized ammonia can travel three hundred miles before settling down on land or water.[20] The smog in much of Southern California contains tiny particles of ammonium nitrate, and the amount that comes from cows may be as great as that which comes from cars.[21]

As for hydrogen sulfide, Robbin Marks, author of the authoritative book *Cesspools of Shame*, notes that it "can cause eye, nose, and throat irritation, diarrhea, hoarseness, sore throat, cough, chest tightness, nasal congestion, heart palpitations, shortness of breath, stress, mood alterations, sudden fatigue, headaches, nausea, sudden loss of consciousness, comas, seizures, and death."[22] Even at less dire concentrations, it stinks like rotten eggs.

The odorless climate-warming gases methane and carbon dioxide can asphyxiate a person if encountered in very high concentrations. Methane, the principal ingredient in natural gas, can also explode.[23] At summer camp, young Denis and his pals set their farts on fire, proving that (1) boys and girls have different concepts of "fun," and that (2) some farts contain methane. (Almost half of humans, however, produce flatus containing no methane, because they lack a single-cell gut organism called archaea. Wondering whether you're a producer? Now you know how to find out.)

Methane is lighter than air, and carbon dioxide is heavier than air, so when air doesn't circulate, methane settles near a ceiling and carbon dioxide pools in a hollow or near the floor. Be wary about entering any enclosed space—or even a pit—where biogases that you can neither see nor smell can accumulate.

Ozone, a corrosive form of oxygen, also forms in lagoons. Up in the stratosphere, ozone protects us from ultraviolet B (UVB) radiation, which can cause skin cancers, cataracts, and weakened immune systems. But when ozone is generated closer to the ground it causes trouble. (Low-down ozone has no connection to the "ozone hole" that appears in the stratosphere.) Ozone is the main component of smog. Breathing it can inflame lungs, impair lung functioning, trigger asthma attacks, and cause coughing, sore throats, and shortness of breath. Young adults seem most vulnerable to ozone's effects,[24] and young adults compose much of the work force at factory farms. Recall how California cows can contribute as much ammonium nitrate to smog as cars do? In some regions, such as the San Joaquin Valley of California (home to both Interstate 5 and the Harris Ranch), cows and their silage also produce more ozone than cars and trucks.[25]

Recently, new concerns about ozone have been getting press. Ozone near the ground is formed in part from methane, and like methane is a short-lived global warming gas. This low-down ozone attacks plants and has reduced food production. Therefore, reducing methane would reduce ozone and might increase crop yield.

Federal clean-air regulations focus on urban areas. Because air-monitoring programs are rare in rural areas, the neighbors of factory farms need more than their noses to prove the existence of pollutants. Concerned researchers collected air samples at twenty houses a quarter-mile or less from large dairy farms in Yakima Valley, Washington, and compared them to samples of air taken at twenty houses three or more miles distant from factory farms. The scientific results confirmed common sense: People in the houses closer to the feedlots were breathing in more ammonia and particulates, including tiny bits of dry cow poop. Ammonia levels were twice as high in houses closer to factory farms. The researchers also found higher levels of Bos d 2, an allergen found only in cows—yes, some people are allergic to cows. Bos d 2 is in cows' dander, urine, and sweat. (Different breeds of cows shed different amounts of Bos d 2.) Indoor air samples of Bos d 2 were ten times higher in houses closer to factory farms. Dairy representatives pointed out that none of the individual pollutants were found in concentrations thought to be unsafe. The principal investigator of this peer-reviewed study, D'Ann Williams, responded that the way pollutants act when combined might make them more dangerous. Both sides agreed more research is needed.[26]

Eventually, federal regulations might limit how much ammonia, ozone, hydrogen sulfide, particulates, and greenhouse gases farms can emit. But don't hold your breath.

Alarming amounts of pollutants from lagoons also get into water. Heavy rainfall, overloading, and breaks in earthen walls cause spills. Such breaches are frequent, and they flood untreated sewage into rivers and estuaries as well as into groundwater. In addition to nitrogen, phosphorus, and bacteria, cow waste often contains natural and synthetic hormones, antibiotics and other drugs, heavy metals, pesticides, herbicides, and salts that can make water unusable for drinking or irrigation.[27]

For example, scientists studying cow pies (possibly not the career path they had envisioned in their youth) learned that large amounts of the hormones given to cows pass right through them. Melengestrol acetate, an artificial growth hormone not approved for human use, can be found in soil for nearly two hundred days after it is injected into a cow.[28] Toxicologist L. Earl Gray Jr. sampled water upstream and downstream from Nebraska feedlots and discovered that male fathead minnows living downstream had shrunken testes and smaller skulls.[29]

Across the nation, determined groups of rural citizens are fighting back against factory farms. *Animal Factory*, by David Kirby, gives blow-by-blow accounts of some of these battles. Family farmers who protest CAFOs have been called terrorists, had their windows shot out, found dead animals on their property, and been threatened by other vehicles on the road.[30] Nevertheless, they've won important battles and were heartened when a 2008 report by the Pew Commission, "Putting Meat on the Table: Industrial Farm Animal Production in America," was published. This in-depth study concluded that CAFOs create unacceptable risks—to cows, the environment, and public health.[31]

"I Caught It from My Cow"

The entangled history of humans and cows is evident in the ease with which we swap diseases. About 91 percent of human genes have been located on cow chromosomes.[32] Our genetic similarity makes it easy for many infections to ping-pong between cows and humans. Runoffs from feedlots are a frequent source of human exposure to some of these infections and the drugs used to prevent or cure them. But first, a look at two diseases of great historical interest that underline the bond between cows and humans.

One of cows' great gifts to mankind was to infect us with a virus: cowpox. Dairymaids caught cowpox by milking cows that had pustules on their udders. The dairymaids got red blisters. At the end of the eighteenth century, a rural English doctor, Edward Jenner, demonstrated that cowpox protected its human victims from getting smallpox (a disease only humans get). Cowpox is much milder than smallpox, which for at least three thousand years had been the scourge of mankind, killing about a third of its victims. The mummy of Ramses V (BC 1150–45) has a pustular rash, which may have been caused by smallpox. Cowpox viruses were used to make the smallpox vaccine, the first safe, successful vaccine for humans. In fact, the word "vaccine" is derived from the Latin *vaccinus*, which means "relating to cows." Because smallpox, unlike cowpox and many other poxes, has no animal hosts except humans, a worldwide program led to its total eradication in 1979. So thank you, cows! Unlike smallpox, cowpox is still around, mostly in the United Kingdom, where it lives on in woodland rodents. It is now rare in cows.

For centuries the greatest threat to humans from cows was tuberculosis. Cows and humans have been swapping tuberculosis (TB) bacteria since the Stone Age. *Mycobacterium tuberculosis* is the strain usually associated with humans, and *Mycobacterium bovis (M. bovis)* is most often associated with cattle (and bison, elk, and deer). But cattle and humans can both get both forms. As recently as 1900, prior to the pasteurization of milk products, TB was second only to pneumonia as a cause of death in the United States. Today 5 to 10 percent of Americans still carry latent TB. Similarly, TB was once the most common disease in cattle here, but control measures have now made it rare, and Americans today seldom catch the cow form of the disease. *M. bovis* is amazingly adaptable, however, and can infect all warm-blooded vertebrates, making it hard to eradicate. In the blunt words of one expert, the best way to control the spread of the disease once it is discovered in cows is to "depopulate the entire herd." This was done after an outbreak in central California in 2008 when 4,800 cows were "depopulated."[33]

TB can be transmitted through the air in invisible droplets, through drinking unpasteurized milk, by eating unpasteurized milk products, or through breaks in the skin. Most infected cows don't show symptoms, so if the infection is discovered, the discovery usually occurs at the slaughterhouse when the lymph glands and organs of cows are inspected.

Imagine you're walking along a rural road and come across a cow with white bubbles foaming around its lips. Is the cow sick with something you could catch? Might the cow bite you?

The cow's frothy mouth could be any of three viral diseases: foot-and-mouth, vesicular stomatitis, or rabies, all of which are occasionally transmitted to humans. Foot-and-mouth disease (FMD) is rarely found in humans, and when it does occur is usually mild. Vesicular stomatitis presents just like foot-and-mouth disease. It makes blisters on cows' mouths and feet and gives the few people who catch it flu-like symptoms and blisters. Rabies, a third possibility, is a greater threat to humans. It might be possible to catch rabies if you drink unpasteurized milk from an infected cow, but the most frequent way ranchers and veterinarians get it is when they think a cow is choking on something and thrust an arm down its throat to grasp the offending object. Although cows rarely bite humans, a cow with rabies might actually attack and bite you. According to the Centers for Disease Control and Prevention (CDC), in 2010, seventy-one rabid cattle were reported in the States, in 2011, there were sixty-five. Because the risk is usually low, cows aren't generally vaccinated against rabies.[34]

Cows can also pass along other viral diseases such as influenza and (in Africa and Arabia) Rift Valley fever.

It's bacterial, not viral, diseases that present the greatest threat from factory farms. Cows crowded together stand withers-to-withers in feces and puddles of urine; their hooves, legs, and udders become caked with filth. Clouds of flies buzz around them. Standing in feces and eating a high-grain diet causes cows' "toes" to grow and curl up, and cattle become lame. An environment that is hell for cows is paradise for germs. Filth on cows' hides and intestines gets mixed in with their meat during slaughter.

Prior to the era of factory farms and pasteurization, bacterial diseases frequently spread from cows to humans. Unpasteurized cow milk was once a common source of typhoid fever, scarlet fever, and diphtheria. Other bacterial diseases that affect both cows and people are anthrax, brucellosis, campylobacteriosis, leptospirosis, listeriosis, and Q fever. Some researchers think the infectious agent that causes Johne's (pronounced YO-knees) disease in dairy cattle might cause Crohn's disease in humans.[35]

Today, however, salmonella is the food-borne bacterium that causes the most hospitalizations and deaths, and factory farms are incubators for some of the most dangerous strains.[36] Symptoms include bloody diarrhea, mucus, and pus. Salmonella outbreaks connected to ground beef occur with depressing frequency. The pathogen is also found in a wide range of other foods and animals.

E. coli is another bacterial pathogen that gets a lot of media attention. Outbreaks and food recalls due to *E. coli* occur year after year. *E. coli* is omnipresent in fecal matter, even in yours. Most of the hundreds of *E. coli* strains are harmless. But several strains found in cow feces are very dangerous because they produce Shiga toxin. Shiga toxins inhibit protein production in cells they infect.

Escherichia coli O157:H7 (sometimes called "hamburger *E. coli*") is Shiga toxin–producing and causes over a third of *E. coli* infections. Infected cows aren't much bothered by O157:H7, but swallowing just a few of these bacteria can sicken people. Diarrhea, vomiting, cramping, and urinary problems are typical symptoms. About 6 percent of victims get hemolytic uremic syndrome. In American children, O157:H7 is the most common cause of kidney infections. In extreme cases the germ causes bloody diarrhea and death. The USDA banned the sale of ground beef containing O157:H7 in 1994, but that hasn't stopped it from being sold. (Here's a shocker: If you catch a Shiga toxin strain of *E. coli* and take most kinds of antibiotics, the antibiotics might kill you. Ciprofloxacin, for example, kills the bacteria, but as they die the microbes can release a lethal amount of toxin.)

Stephanie Smith, a young woman who once taught dancing to children, is one of O157:H7's many victims. Eating a hamburger gave Stephanie hemolytic uremic syndrome. Her brain and kidneys were damaged; she suffered seizures, was in a coma for nine weeks, and became paralyzed from the waist down. As reported in a Pulitzer Prize-winning story by Michael Moss in the *New York Times*, the patty Stephanie ate was actually a mixture of scraps and trimmings from four slaughterhouses. Over two dozen other people who ate the patties also got sick. Unlike Stephanie, nearly all of them eventually recovered.[37]

In Stephanie's case, the global food, agricultural, and financial corporation that had made the hamburgers, Cargill, had largely relied on its suppliers to check for dangerous bacteria; it did its own check only after the ingredients were ground together, an approach said to be less likely to find contamination. Reporter Michael Moss was told by two grinding company officials that because of their suppliers' fear of expensive, business-busting recalls, many big slaughterhouses won't sell to grinders who plan to test their shipments for *E. coli* before grinding and mixing them with others.[38]

Feeding a high-grain ration to cows prior to slaughter, which is the practice at factory farms, increases the amount of O157:H7 in their feces.[39] A 2000 survey found the prevalence of *E. coli* O157:H7 in the pre-evisceration carcasses of cows was much higher than previously thought: 83 percent. This fell to a still alarming 17 percent in post-processing samples from the cooler.[40] Massive amounts of meat and vegetables are recalled (and wasted) in the United States because they are contaminated by *E. coli* O157:H7. A California firm, Huntington Meat Packing, recalled 864,000 pounds of ground beef in January 2010. Later that same year, Valley Meat Co. of Modesto, California, recalled a million pounds of O157:H7-contaminated ground beef.[41]

Mysteriously, in 2010 Nestlé (the largest food company in the world) found *E. coli* O157:H7 in some batches of its cookie dough, a substance that a surprising number of people like to eat raw. Thirty-four raw-dough nibblers were hospitalized, but none died. How did bacteria associated with cows get into cookie dough? None of dough's ingredients usually host *E. coli*. One theory is that the bacteria was in the flour. Wheat contaminated with harmless *E. coli* and with salmonella has been found, so wheat can probably be contaminated with the dangerous sort of *E. coli* as well.[42]

Because so much of a cow's meat (in the neighborhood of 40 percent) is left over after named cuts have been removed, the scraps are saved and ground up. Meat from hundreds or even thousands of cows may be contained in a single lot of ground beef.[43] This mixing greatly increases a consumer's chance of swallowing toxic bacteria, antibiotic-resistant bacteria, or antibiotic residue. As of June 2013, one out of every 354 samples of raw ground beef the USDA had tested for *E. coli* O157:H7 were positive. (The testing program began in 1994.) Even more worrisome, one out of 118 was positive for one of six other types of non-O157 Shiga toxin–producing *E. coli.*[44]

In June 2012, over the outraged bellows of the beef industry, the USDA expanded its monitoring program to include searches for six "rare" types of Shiga toxin–producing *E. coli.* The Centers for Disease Control and Prevention estimates that these strains cause around 173,000 infections every year.[45] Although ground beef containing them can no longer be legally sold, the bans are sometimes probably ignored, as they were with O157:H7. Furthermore, the new bans apply only to ground beef, and not to all beef products, nor do they apply to vegetables (the source of most such infections). Even the ground beef ban doesn't apply to other infectious bacteria, or to all dangerous strains of *E. coli.*[46]

The mechanical tenderizing processes used by most beef producers allow cross-contamination to occur even in steaks. The CDC has attributed at least three *E. coli* outbreaks to this practice.[47] To tenderize meat, a conveyor moves hunks of meat under a device with many vertical needles or blades. The device isn't washed between injections, so any germs in one hunk of meat can be spread to many. All the big four beef packers (JBS Beef, Tyson Foods, Cargill Meat, and National Beef Packing) have tenderizing machines. These corporations process 80 percent of all beef slaughtered in the United States.[48]

Cooking can kill bacteria and viruses, but some folks pride themselves on how bloody they like their rare steaks—"just so it doesn't trot off my plate"—and therefore can get sick from eating steak.

Shoppers can't tell if packaged meat has been mechanically tenderized just by looking at it, so consumer advocates have long begged for labels that reveal whether meat has been tenderized. Mike McGraw highlighted this issue in an excellent series of articles in the *Kansas City Star* about "Beef's Raw Edges."[49] In 2013, the USDA proposed rules to require labels that told whether meat had been mechanically tenderized; the American Meat Institute (a trade group) said labels would only confuse shoppers.[50]

Now some good news: Although 2010 saw many recalls, in June 2011 the CDC reported that illness due to food-borne *E. coli* O157:H7 had declined to CDC's goal of about one case per one hundred thousand people, a drop of about 50 percent since 1997. They attribute this decline to PulseNet (a national system of eighty-seven laboratories created in 1996 to detect pathogens), cleaner slaughtering methods, better inspection of ground beef processing plants, 234 beef recalls since 1994, and less undercooking of beef.[51] The federal government's success with *E-coli* O157:H7 shows it is possible to improve food safety.

***E. coli* don't originate in vegetable fields.** Their natural habitat is the lower intestines of warm-blooded organisms. Salmonella is another gastrointestinal disease from animals. But once these bacteria exit an animal, in poop, they can live for quite a while in the environment, where they contaminate vegetables. In 2010, lettuce contaminated with toxic *E. coli* was recalled in twenty-three states.[52] From 1966 to 2011, at least thirty outbreaks of salmonella and *E. coli* were linked to raw or lightly cooked sprouts.[53] *E. coli* O157:H7 has been found in spinach and lettuce irrigated with water downstream from cattle.[54] *Consumer Reports* even found fecal contamination on "prewashed" salad greens.[55] In 2006, antibiotic-resistant *E. coli* O157 on spinach sickened nearly two hundred people living in twenty-six states. Three died. The following month, the same strain of resistant bacteria was found on a cattle ranch bordering the spinach field.[56]

We were shocked to learn that pathogens such as O157:H7 can actually be sucked inside individual leaves of lettuce and spinach, where they can't be washed off—and nobody cooks lettuce.[57] Broccoli, in contrast, contains substances that appear to suppress O157:H7 infections.[58]

In addition to viruses and bacteria, protozoa and fungi from cows infect humans. Protozoa (single-cell creatures that have a nucleus and can move around under their own power) usually travel from cows to people in water, but they can also be spread through contact with cowpats. We learned a bit about *Giardia*, a protozoan parasite that causes giardiasis, when we lived on Lookout Mountain, above Golden, Colorado. Warm-blooded animals like cows and beavers carry *Giardia*. Giardiasis is sometimes called beaver fever, because people can catch it by drinking water downstream from a beaver dam. Our reservoir was called Beaver Brook Reservoir. Enough said. Water from it was then untreated, so we had an elaborate system of filters in our basement.

Protozoa such as *Giardia* and *Cryptosporidium* thrive in feedlots. If they get into water or food they can cause diarrhea, fever, kidney failure, and occasionally death. Spreading *fresh* manure on fields can contaminate both water and crops.

Even some fungi bounce between humans and cows. Ringworm (which is spread by spores, not worms) is a common skin disease in cows and especially prevalent where cows are crowded together. Athlete's foot and jock itch in humans are ringworm. Ringworm spreads through direct contact with an infected animal.[59]

The Problem with Denis's Elbow

One of medicine's greatest advances—the use of antibiotics to fight infections—is rapidly being eroded by the advent of resistant "superbugs." Because so little is being done by most governments and drug companies to address this problem, the director general of the World Health Organization, Margaret Chan, recently warned that we might be entering a post-antibiotic era, when even a scraped knee might lead to death."[60] Or a scraped elbow, as we learned firsthand.

One night in 2009, Denis went for a run. He tripped over a tree root and fell forward, scraping his right elbow. The injury was minor and, being male, he forgot about it. A few days later Gail noticed his elbow was as swollen, shiny, and red as Rudolph's nose. Being female, she urged him to see a doctor.

The doctor said it was staph and prescribed an antibiotic. The drug didn't work. He prescribed another. The elbow continued to swell and redden. Denis was then diagnosed with the dreaded MRSA: methicillin-resistant *Staphylococcus aureus.* Left untreated, some strains of MRSA destroy vital organs, cause toxic shock syndrome, and open the door to flesh-eating diseases and death. MRSA are resistant not just to methicillin but to penicillin, oxacillin, amoxicillin, and some other antibiotics.

Denis's doctor eventually found a drug that worked. But it still took months for the redness to fade and the elbow to shrink back to normal size. The CDC estimates that in 2010 there were 82,040 infections and 11,478 deaths from MRSA. Antibiotic-resistant bacteria of all kinds make an estimated two million Americans sick every year; 23,000 of them die from the infection (the CDC intentionally underestimates the numbers, counting only those deaths that directly result from the resistant bacteria).[61] The CDC estimates that antibiotic resistance adds $20 billion a year to America's health-care bill. The Center for Science in the Public Interest unearthed thirty-eight outbreaks from 1973 to 2009. Most were linked to dairy products and ground beef. Disturbingly, nearly half occurred in the last eleven of the thirty-eight years studied.[62]

The more frequently people and animals are exposed to antibiotics, the faster resistance develops. Bacteria swap genes with the eagerness of people at a convention swapping business cards. Just by bumping into each other and touching cell walls, they can trade genes. This ongoing, worldwide swap meet has resulted in superbugs resistant to multiple antibiotics.

Bacteria reproduce by splitting in two. Streptococci can divide every twenty minutes. After one hour there are eight; after four hours, there are 4,096. In eight hours, 16,777,216. With each new generation, a mutation can occur that will allow a bacterium to survive better than its fellows. Genetic mutations are passed along to progeny. Therefore, a bacterium resistant to an antibiotic present in its environment quickly outmultiplies other bacteria. Giving animals low doses of antibiotics turbo-charges the development of resistance.[63] When an antibiotic is administered in doses too low to kill all the targeted bacteria, those bacteria with any level of antibiotic resistance live on, multiply, and can further mutate into forms that are even more resistant.[64]

MRSA is only one of the superbugs springing up in hospitals and communities. Diseases with strains resistant to one or more families of antibiotics now include anthrax, gonorrhea, strep infections, pneumonia, staph infections, meningitis, shigellosis, tuberculosis, typhoid fever, and urinary tract infections. Meet *Acinetobacter baumannii*, or MDRAB. Nicknamed "Iraqibacter" because of its prevalence in American troops sent there, some experts think it's a bigger threat than MRSA.[65] Then there's *Klebsiella pneumoniae*, a bug found in hospital settings. Like MDRAB, it is resistant to virtually all of today's antibiotics.

Although the CDC's National Antimicrobial Resistance Monitoring System (NARMS) surveys retail meat looking for multidrug-resistant *Campylobacter*, *Salmonella*, *Enterococcus*, and *E. coli* (all four are bacteria found in cow intestines), it doesn't look for other pathogens that have become resistant. But it should. Consider *Staphylococcus aureus*, which infected Denis's elbow. In 2011, The Translational Genomics Research Institute did the first nationwide survey to find out how pervasive *Staphylococcus aureus* is in ground beef, pork, and poultry. The researchers looked at 136 samples representing 80 brands from 26 grocery stores in 5 cities.[66] About half of all samples were contaminated with *Staphylococcus aureus*, including 37 percent of the ground beef samples. One-third of the contaminated samples had a strain of *Staphylococcus aureus* resistant to three or more antimicrobials. Various strains displayed resistance to tetracycline, ampicillin, penicillin, erythromycin, fluoroquinolones, quinupristin/dalfopristin, oxacillin, daptomycin, and vancomycin. When this story broke, the American Meat Institute, which represents meat processors, correctly noted the limited sample size and reminded readers that proper cooking kills microbes.[67] "Proper" means well done, not rare, and certainly not steak tartare.

Cows' Role in Antibiotic Resistance

Crowding cows into factory farms and feeding them high-grain diets results in filthy living conditions and mental and physical stress. This causes cows to get sick. As noted earlier, factory farms routinely put low doses of antibiotics in cows' food or water to prevent illness. With dairy cows producing copious amounts of milk, far more than they did in Grandpa's day, they are similarly stressed and more vulnerable to disease.

Eighty percent of all antibiotics sold for all purposes in America are given to healthy animals in factory farms to improve feed efficiency, prophylactically guard against disease, and promote growth.[68] In a 2001 report, *Hogging It!: Estimates of Antimicrobial Abuse in Livestock,* the Union of Concerned Scientists estimated that 3.7 million pounds of antibiotics were being given to healthy cows every year. In early October 2014, the Food and Drug Administration (FDA) estimated that the use of medically important antibiotics in animals increased by 16 percent from 2009 to 2012.

Food animals are given antibiotics mixed in with their feed or drinking water. These drugs include many antibiotics that people rely on, familiar medicines like cephalosporins, penicillin, tetracycline, and erythromycin. The FDA did an analysis covering the first decade of the new millennium and found that most of the antibiotics given animals were probably contributing to the estimated annual 23,000 human deaths and 2 million human illnesses caused by resistant bacteria. It took a Freedom of Information Act request by the Natural Resources Defense Council to uncover this information.[69] Although food industry officials point the finger of blame at hospitals, and nearly everyone agrees that the overuse of antibiotics by the public is also partly responsible, most scientists think that continually giving factory-farm animals antibiotics contributes greatly to antibiotic resistance.

It's still a mystery why healthy animals gain weight more quickly when fed antibiotics. Human infants given therapeutic-strength antibiotics also pack on the pounds. A longitudinal study of 11,532 children discovered that infants treated with antibiotics when they were less than six months old were 22 percent more likely to be overweight at age three than children who weren't exposed to the drugs. Scratching your head? So are scientists. Maybe, they postulate, antibiotics change the composition of the microbial stew in a baby's body, which leads to a weight gain.[70] Maybe the same applies to animals. In the 1950s, an outrageous experiment was conducted on spastic, mentally challenged children. Those given antibiotics twice daily gained more weight. Similar weight gains were seen in navy recruits given antibiotics for seven weeks. In an opinion piece for the *New York Times*, Pagan Kennedy asked whether all the antibiotic pills we give our children today (along with the smaller amounts of residual antibiotics in food) could be promoting an increase in body fat.[71]

We were surprised to learn that antibiotics even get into cows as a side effect of growing corn-derived ethanol. Dried distillers grain (DDG)—a low-starch, high-protein by-product of ethanol production—is a popular cow supplement at feedlots. The ethanol mash from which DDG is made is often contaminated with *Lactobacilli*, which thrive in the warm, moist, sugar-rich environment. The *Lactobacilli* compete with alcohol-producing yeast for the sugars, but they produce lactic acid, not ethanol. To combat the *Lactobacilli*, many ethanol producers add antibiotics, including penicillin and erythromycin, to their fermentation tanks. After the ethanol is removed, the tanks still hold distillers' grains, which typically contain unregulated antibiotics. The Institute for Agricultural Trade and Policy argues persuasively that pharmaceutical companies selling antibiotics for this purpose are violating the law.[72]

Many respected organizations have called for a ban on giving healthy farm animals antibiotics that are also used in humans. They include the World Health Organization, the Centers for Disease Control and Prevention, the American Medical Association, the American Society for Microbiology, the American Public Health Association, the Union of Concerned Scientists, the American Academy of Pediatrics, and the American Pharmacists Association.[73]

In 2006 the European Union banned the use of antibiotics to promote growth in farm animals. They had already banned giving farm animals antibiotics used on humans.[74] Denmark has gone further and prohibits veterinarians from making money by selling farmers antibiotics (as vets can do in the United States), eliminating this incentive to prescribe unnecessary drugs.[75] Danes now use less than half the antibiotics on farms they once did, with no bad consequences.

Astonishingly, American dairy farmers and ranchers still don't even have to get a prescription to buy antibiotics for cows—they can just pick up the drugs at a feed store. And American veterinarians can still prescribe "extra-label" uses of most approved veterinary drugs. This means a drug can be used for a different species, for a different indication, at a different dose, or administered differently than it was approved for.[76] The FDA recently, and reluctantly, prohibited the extra-label use on farm animals of a class of antibiotics called cephalosporins (often given people who have strep throat, pneumonia, urinary tract infections, and skin infections). This means cephalosporins can no longer be used to prevent cows from getting sick or given to cows in doses too low to cure disease.[77] Unfortunately, the FDA only acted in response to a lawsuit by the Natural Resources Defense Council.[78] According to NRDC: "Revealingly, in court documents, the agency said that if it started limiting antibiotics in animal feed, the livestock and pharmaceutical industries would protest so strongly it would consume too much of the agency's time and resources, suggesting the agency is acting on its fear of the pharmaceutical and agricultural industries rather than doing what is best for human health."[79] Georgetown law professor Lisa Heinzerling argues convincingly that the FDA had all the power it needed to protect public health in a timely manner without going through the years-long formal hearings process but was immobilized by "habits of mind that entrench agency inaction."[80]

In March 2012, a federal magistrate in New York said the plaintiffs (NRDC) had won their case on the undisputed facts presented, so there was no need for a trial. He ordered the FDA to inform drug manufacturers that penicillin and tetracycline will also soon be banned to promote growth in animals. Unfortunately, his order did not include antibiotics used to *prevent* disease. In the whack-a-mole world of antibiotic abuse, all future use of antibiotics to stimulate growth might now be justified as preventing disease.[81]

At the end of 2013 the FDA rolled out a nonbinding proposal to be put into effect over the next few years. It applies to antibiotics also used to treat humans and requires a veterinarian's prescription to get antibiotics for food-producing animals.[82] The use of antibiotics to promote animal growth would be banned. However, it will remain legal under this regulation to give farm animals antibiotics to *prevent* a specified disease in specified animals, if there are no reasonable alternatives. Drug companies are asked to voluntarily revise their labels accordingly. Once they do this, extra-label uses would be illegal. A proposed rule concerning this was available for public comment when our book went to press. Tellingly, an association representing firms that make drugs for animals, and the big drug maker Zoetis, both support the new policy; Zoetis doesn't expect a big impact on its profits.[83] This response suggests that the drug companies don't think the new law will make a dent in their sales by greatly reducing the amount of antibiotics given healthy animals.

U.S. Representative Louise Slaughter's proposed Preservation of Antibiotics for Medical Treatment Act is much stricter and has been endorsed by 368 worthy organizations but by none of her Republican colleagues. The act would restrict the use of antibiotics for disease-prevention or to promote growth (they could still be given to sick animals). Under the proposed law, drug companies would need to demonstrate that giving animals existing or new antibiotics critical to humans would not lead to the development of resistance. As of September 2013—in an oblique tribute to the power of the agribusiness lobby—GovTrack.us (an independent, for-profit service supported by ads) gives the act a 3 percent chance of getting through committee and a 1 percent chance of passing the House. If a radically different Congress is seated and the bill does pass, it would provide great benefits for humans. It might also force a dramatic improvement in the lives of cows, because factory farming would become impractical.[84]

Antibiotic Residues: From Cows to You

Because up to 80 percent of antibiotics given to cows orally pass through them intact,[85] antibiotics in manure get into our soil, surface water, and groundwater.[86] From there they get into farmworkers, drinking water, and the general public.[87]

Antibiotics are not benign. Bad reactions to antibiotics send over 140,000 Americans to emergency rooms each year.[88] Flunixin, for example, causes gastrointestinal erosions and ulcers and the death of kidney cells. Ivermectin kills neurons. The side effects of fluoroquinolones (Cipro, Levaquin) can be so nasty—examples include ruptured Achilles tendons—that they must carry a "black box" warning. Even low doses of any antibiotic can mess up your intestinal biota.

When meat that contains antibiotics gets into the food supply, it causes problems in sensitive people. Reactions to residues of common antibiotics (for example, penicillin) in meat can even send people to the hospital, although such severe reactions are uncommon.[89] To try to keep antibiotic residue out of beef and milk, federal regulations require cattle to be taken off antibiotics a specified number of days prior to slaughter; milk with antibiotic residue should be discarded. Unfortunately, rules aren't always followed. Furthermore (and somewhat inconsistently), the FDA allows what it considers safe levels of antibiotics in milk. These levels are not set low enough to protect all sensitive people.[90]

The meat of former dairy cows, which constitutes roughly a fifth of our meat supply, is more likely than most beef to contain harmful antibiotic residues.[91] The male calves of dairy cows often quickly become veal chops. But before they are killed, these calves may be given "waste milk" from cows on antibiotics, milk that can't be sold for human consumption. They are also given medicated feed. Thus, veal is more likely than most beef to contain drug residues. In 2008, over 90 percent of beef contaminated by antibiotic residues came from plants processing dairy cows and veal. Common sense suggests such plants deserve stricter scrutiny, a 2010 Office of Inspector General report said. But the plants don't get it. Recall of such tainted meat is voluntary, and the USDA's Food Safety and Inspection Service (FSIS) hasn't requested any such recall since 1979.[92]

The National Conference on Interstate Milk Shipments, an industry group that declares its goal to be "assuring the safest possible milk supply for all people," passed a resolution to stop the FDA from ordering additional drug tests on dairy farms that provided tainted veal. The Center for Science in the Public Interest (CSPI) urged the FDA not to go along with this.[93] The FDA sided with CSPI on this issue.[94] When the agency tried to implement its plan in 2011, however, the Dairy Farmers of America (the nation's largest dairy coop) and other dairy-industry supporters angrily protested, and the FDA huddled down with them for more consulting.[95] The agency's final plan hides which farms or states tainted tested milk comes from, and it is for informational purposes only.[96]

Also in cow poop are any nasty antibiotic-resistant bacteria created by the misuse of antibiotics. Neither storing waste in a lagoon nor using biofilters to clean waste is an effective method of destroying bacteria. Anaerobic digestion (processes where microorganisms digest material in an oxygen-free environment) and composting at moderate temperature do greatly reduce the number of antibiotic-resistant bacteria in animal manure.[97] In 2006, however, the EPA decided that it wasn't cost reasonable to require factory farms to implement such practices to control pathogens.

Alternatives to antibiotics are needed. Before antibiotics came along, immunoglobulins from bovine colostrum—the thick yellow fluid that mammal mothers make for their newborns—were used to fight infection. Today, as more disease-causing bacteria become resistant to antibiotics, researchers are taking a second look at cow colostrum. Produced for a couple days after a baby is born, colostrum bolsters a newborn's immune system while providing nutrition. It is difficult to synthesize. Although bovine colostrum differs from human colostrum in the concentrations of key immunoglobulins, it might contain immunoglobulins, cytokines, and other bioactive components that could prevent or treat infections in humans.[98]

Cows that have been vaccinated against particular diseases and developed antibodies to fight them produce "hyperimmune" bovine colostrum. In clinical trials, bovine colostrum has helped babies with certain infections and adults who have undergone open-heart surgery or surgery for gastrointestinal problems, as well as AIDS patients with diarrhea. So although cows contribute to the antibiotic resistance and residue problems, these useful animals might also be part of the solution.

You can nearly always protect yourself from pathogens, even superbugs, by cooking beef to 155 degrees Fahrenheit (160 degrees Fahrenheit for ground beef), by avoiding unpasteurized dairy products, and by cooking vegetables. Unlike bacteria, however, many antibiotics are not destroyed or rendered ineffective by cooking. It is also possible that toxic residues form when some antibiotics are broken down by heat.[99] Some antibiotics are heat-sensitive and may be destroyed by pasteurization; others aren't.[100] You can help to protect your family (and our society) from antibiotic residues and resistant bacteria by choosing organic food. A big advantage of products from organic cows is that they are free of antibiotics and do not put these drugs into the environment or you.

Beware labels that say "antibiotic-free" and "no antibiotic residues." They mean only that the seller claims that the packaged beef contains no antibiotics. Even if this is true, antibiotic resistance develops back in the feedlot; a cow can be treated with antibiotics for years and treatment stopped during just the last couple weeks of its life. "No antibiotic growth promotants" means that the feedlot operator alleges that any antibiotics used were meant to prevent disease, not promote growth. But they are the same antibiotics, whatever the cattleman's ostensible intent. The USDA does, however, approve labels such as, "never ever given antibiotics" or "no antibiotics ever." But also look for a "USDA process verified" logo if you see such words. The logo means that the USDA has investigated the company's process (not their product) and found the claim to be true.[101]

Field Trip: Poop into Power

Pioneers used buffalo dung in campfires and for cooking. Today, because of the extreme quantities of bovine waste available, a higher-tech method than campfires is needed. For over forty years a few small, farm-based digesters have been turning cow poop into methane. The methane is then burned to make electricity. Recently, much larger digesters have been put to work. The new systems now make more economic sense, because there is a market for alternative energy credits. Grant money is also available to support such efforts.

To be clear: Digesters are more about controlling pollution than generating electricity. If every ounce of manure from 93 million cows were converted to biogas and used to generate electricity, it would produce less than 3 percent of the electricity Americans currently use.[102] If, however, the biogas replaced coal, and if machinery is carefully plumbed to avoid methane leaks, digesters could reduce our greenhouse gas emissions from electricity by about 6 percent. Utilities will pay a premium for such green power.

The "Cow Power" program, run by the Central Vermont Public Service Corporation (CVPS), is a leader in turning poop into power. CVPS helps farmers set up anaerobic digesters and generators. More than 4,600 CVPS customers have volunteered to pay an additional four cents per kilowatt-hour for some, or all, of their electricity to support this green power. If Congress ever places a price on carbon, as would be required by Senator Maria Cantwell's CLEAR Act, market forces would make cow power economical without voluntary subsidies. In Vermont, with the additional sale of waste heat and cow bedding as by-products, it already makes economic sense.[103]

In California, a 2011 law requires utilities to obtain one-third of their electricity from renewable sources by 2020. Subsequent laws require utilities to obtain at least 250 megawatts from biogas. Pacific Gas and Electric Company (PG&E), a San Francisco-based utility, has an agreement with the Huckabay Ridge Anaerobic Digestion Facility to convert poop from Huckabay's ten thousand cows into biogas and put it in a natural gas pipeline. It is the largest anaerobic digestion plant in North America. What makes the deal even more interesting is that Huckabay Ridge is located in . . . Texas.

PG&E won a ruling from the California Utility Commission that the gas pipeline system should be treated much like an electrical grid. It's impossible to trace electrons through a grid; if a wind turbine puts "green" electrons into a grid and some customers in a particular utility district pay a little extra for green electricity, those electrons are attributed to those customers. But for customers' willingness to pay a little more, the wind project would not have been built. Similarly, the gas produced in Texas goes into a pipeline. There is no direct pipeline route from Huckabay to PG&E power plants, so it is unlikely that even one gas molecule from there will wind up at a PG&E facility.[104] But the molecules will displace fossil-fuel-derived natural gas molecules somewhere, and that saving would not have happened had the initiative not been supported by PG&E.

This logic escapes some California consumers who are paying higher prices to create jobs and economic growth in Texas. Because new laws now make biogas production more attractive in California, PG&E says that it plans to get more biogas from its own state in the future.[105]

Digesters are also feasible on a neighborhood scale. Our home state, Washington, is a dairy state. In Skagit and Whatcom counties alone, some sixty-five thousand cows produce about 3.5 *billion* pounds of manure a year.[106] It takes a brave man to dig into that mess. One such man is Kevin Maas. Along with his brother Daryl,

Kevin created Farm Power, which now operates in Washington and Oregon.

On a pleasant summer day in 2011, we met up with Kevin at the Rexville Grocery in Skagit County. "Grocery" doesn't do justice to this rural meeting place with a café as well as shelves of supplies. A tall, tan, muscular man with a fine head of short brown hair and a goatee, Kevin joined us for coffee. The drinks came in small, medium, and large cups—not grande, venti, and trenta. The coffee was excellent.

Growing up, the Maas brothers saw farm family after farm family lose its land because it couldn't compete with factory farms. Friends moved away and community bonds frayed. These were mostly people who had enjoyed working with animals, who had known their cows individually and cared about them. "In my dream world," Kevin said, "people who want to work with animals would be able to do so where they grew up."

It took Kevin three years of planning and learning about business before he was ready to start. A former teacher, he went back to school to earn an MBA from the Bainbridge Graduate Institute, which specializes in sustainable businesses. After that, it took four more years of effort, but the Maas brothers are succeeding. An early challenge was to convince a couple of dairy farmers to do things a little differently. Dairy farmers, like their cows, are most comfortable doing things the way they're used to, ways proven to work. But these farmers knew the Maas brothers and trusted them. Besides, Farm Power was assuming the risk. All the farms had to do was provide free manure.

If a conventional dairy farmer wants to live a middle-class life today, Kevin says, he probably needs at least 350 cows. (Organic dairy farms can succeed with fewer cows.) One conventional farm they work with in Skagit County has 900 cows; an adjacent farm has

Farm Power's biogas system converts dung into electricity, heat, and cow bedding. The digester is in the low mound on the left, the dumping tank far right, and the generator is inside the building. Photo: Gail Boyer Hayes.

350. These two dairies are "scrape farms"—they scrape up manure rather than flushing it out with water. The cows eat mostly roughage (grass and corn stalks), not grain. It's not feasible to pasture cows year-round here because it gets too cold and the pastures get muddy. The 900 cows we visited were in an unwalled barn with lots of air circulation.

The Maas brothers decided to set up their Farm Power plant right between the dairies, so the manure wouldn't need to be trucked long distances to the digester, and the finished product could be piped at reasonable cost to nearby fields. With the farmers lined up, all Farm Power had to do was find $3 million to build a million-gallon tank in which to digest manure, a generator, and tanks to hold the stuff coming in and going out of the digester.

Arranging financing and dealing with regulations took tenacity and creativity. One problem was that banks usually make farming loans only when they get valuable collateral to secure the loan. Farm Power doesn't own farmland, and banks don't consider large digesters collateral. But when the brothers could show that they had a ten-year contract to sell the electricity they would generate to Puget Sound Energy for 7½ cents per kilowatt-hour, the bank finally agreed. (Electricity prices are very low in the Pacific Northwest, thanks to hydropower.) It took a couple more years to work out carbon-offset deals that improved the financial returns and a USDA federal loan guarantee. Farm Power also turned to private investors to raise $90,000.

Our coffee cups drained, we left the Rexville Grocery and Kevin took us to the Farm Power site in Skagit County to see the action. The first thing we noticed was something missing: stinky air. We climbed up on top of the concrete-covered digester tank and watched as a truck full of manure and food scraps dumped its load into an octagonal tank. The truckload of fresh waste stank, but not badly and only briefly. The permit Farm Power got from the state Department of Ecology (after convincing the Department to alter its regulations) allows it to include up to 30 percent pre-consumer food waste—things like cow blood, dead chickens, and fish waste. Food that has not already been digested by animals contains more energy, allowing the anerobic bacteria in the digester to pump out more methane. The facility can process forty to fifty thousand gallons of manure daily.

From the octagonal tank the slew is pumped under pressure into the digester. There are no moving parts inside the digester. Kevin suggested we think of it as a gigantic concrete cow's stomach. The sludge slowly moves from one side to the other. First comes the "acid chamber," where bacteria in the manure produce acid that breaks the manure down into volatile fatty acids and acetic acid. In the second chamber, bacteria turn the volatile fatty acids into methane. The methane gas collects on top, where it is fed into a pipe.

Kevin Maas (left) and Denis admire the Farm Power generator that turns out a profit as well as electricity. Photo: Gail Boyer Hayes.

The generator—a large red noisy machine kept shiny as a fire truck—is housed in its own building. It's bursting with valves and attachments. The generator had been running for two years and would repay the bank loan in another eight years. Together, this generator and another, which Farm Power operates at Lynden, Washington, generate enough electricity to power a thousand homes.

Generators also produce a lot of heat. The heat from this one is fed into the digester to keep it a steady 100 degrees Fahrenheit. That encourages the bacteria to produce methane. There's still heat left over, and Kevin is trying to talk a farmer into putting up a greenhouse to use that heat. Such a greenhouse already exists at the project in Lynden, and the grower pays a modest amount for the heat. The generator runs 24/7, continuously producing 750 kilowatts of electricity.

The liquid material coming out of the digester is a better fertilizer than raw manure because it contains far fewer pathogens and weed seeds and doesn't stink as much. It first flows into a pit; from there, as a more stable manure slurry, it's piped to nearby fields where it can be pumped through an irrigation nozzle or injected into the soil.

The dry residue is turned into sanitary, comfy cow bedding. After the dry matter is squeezed through a screen, it's loaded into trucks and hauled back to the farms. In the future, Farm Power plans to pasteurize the bedding product. Kevin scooped up some finished product stored at one of the nearby dairies. He held it out, inviting Denis to examine it. The bedding was still hot, and smelled like soil and hay.

Once a digester project is up and running it takes only one part-time employee, Ben, to keep an eye on the generator and stir the muck when necessary. The two farms combined save about $100,000 a year on bedding and a total of about $1million a year on all expenses (they get free fertilizer as well as bedding). Putting methane to a good use, rather than letting it contribute to global warming, makes digesters a sweet deal for us all.[107] Farm Power reinvests most of its profits in growing the business, but its investors did get a small dividend.

Digesters don't solve every environmental problem. Certain anti-

Kevin Maas grabs a handful of cow bedding made from residue left over after extracting electricity, fertilizer, and heat from cow manure. Such generators are a win-win idea. Photo: Gail Boyer Hayes.

biotics in cow manure can kill off the fermenting and methanogenic bacteria that make the process possible.[108] The heat in digesters probably doesn't destroy most antibiotics. New research suggests some pathogenic and antibiotic-resistant bacteria survive anaerobic digestion.[109] Installing a scrubber to remove sulfur dioxide from the digester gas wasn't economically feasible for the Maas brothers, so they got a permit to emit some pollution. More nitrogen, phosphorus, and potassium remain in the final product than is ideal. Carbon dioxide is also put in the air, and the trucks hauling waste and bedding burn fuel.

Even when biogas is scaled up to the PG&E level, aside from PG&E's ability to scrub out sulfur dioxide, all the environmental problems with biogas encountered by the Maas brothers remain. Nevertheless, the total warming gases emitted by the digesting process add up to only a small fraction of what dairy-farming-as-usual contributes. Although not a panacea, bioenergy is a good step in the right direction.

Another, higher-tech approach to reducing the warming gas methane in dairy waste is to turn it into plastic. Two young men have been diligently working on a project to do this. Their product, AirCarbon, can be used to make chairs, storage containers, and other items. Eventually, if it replaces a great deal of plastic made from oil, it could have a noticeable impact. The 2013 International Conference on Bio-based Plastics and Composites gave it their "bio-material of the year" award. Like making electricity from dairy waste and fine-tuning cows' diets, this clever idea might mitigate, but certainly won't eliminate, 93 million cows' contribution to warming. Nor will it solve the problem of where to put the megatons of hazardous waste that fill the nation's tens of thousands of lagoons.

Chapter Five

GOT MILKED?

Cow milk is one of the most hotly debated topics in nutrition. Surprised? We were. We learned that much conventional wisdom about milk is not backed up by solid scientific evidence.

The milk lobby has bombarded Americans with the message that consuming milk is healthy: "Every Body Needs Milk," "Milk Has Something for Everybody," "Milk Your Diet, Lose Weight," "Drinka Pinta Day," "Milk is the Perfect Food," "Body by Milk," "Milk Is a Natural," and the two-decades-long "Got Milk?" featuring celebrities with milk mustaches. Worried about slumping sales, in 2014 the Milk Processor Education Program announced a new campaign: "Milk Life."

The National School Lunch Program requires schools to provide nonfat or low-fat milk. Federal food assistance is also heavy on milk and cheese. In 2011 a typical American swallowed 604 pounds of dairy products.[1] But the USDA reports that between 1975 and 2012 per capita fluid milk consumption fell by a quarter. Each new generation consumes less milk—even if race and income are accounted for. The USDA worries that this decline in milk consumption will worsen Americans' health.[2]

But is cow milk truly good for everyone? We dove into milk and emerged coated with serious doubts. Some of what we learned was serious enough to cause us to change our consumption habits: For reasons explained below, we now avoid nonorganic dairy products, and Denis has greatly reduced his milk intake.

A mother's milk is amazing: human or cow, it comes customized for her particular baby. A human baby's gender is taken into account in terms of how much fat and protein are in a mother's milk. And human milk is further adjusted according to whether times are good. For example, an affluent, well-nourished woman will provide more fat and protein in her milk for a male infant. When times are lean, a girl baby gets more fat.[3] Why might this be? The Trivers-Willard hypotheses (proposed by Robert Trivers and Dan Willard) speculates that (especially in polygamous societies) if times are lean, the condition of a mother might be poorer. A son born in lean times may be unable to afford wives. But a poor daughter might still find a husband. So a girl would then be the better bet to provide grandchildren. This theory also applies to polygamous animal species. More recently, Melissa Larson and others have proposed that the level of glucose circulating in a cow mother's blood is responsible—the higher the level (high levels are more likely when the mother's condition is good), the more likely a male blastocyst will survive.[4]

A Holstein cow produces more milk if she has a female calf, and even more milk if she births two daughters in a row.[5] Why is still a mystery. Cow milk differs from human milk: It's richer and less sweet. Calves grow fast, so cow's milk has more protein, sodium, potassium, phosphorus, and chloride than is good for a human baby, and not enough unsaturated fats, essential fatty acids, iron, retinol, and vitamins E, C, and D. Even hormones have been shown to vary in animal milk according to the sex of the baby[6] (It seems possible hormones may vary by species, too.)

Dairy products are nutritious and can be part of a healthy diet for many children and adults. But dairy's downside is usually downplayed or outright ignored. Walter C. Willett, Chairman of the

Department of Nutrition at the Harvard School of Public Health, provides this counterbalance to the industry's claims:

> [T]he recommendation for three servings of milk per day is not justified and is likely to cause harm to some people. The primary justification is bone health and reduction of fractures. However, prospective studies and randomized trials have consistently shown no relation between milk intake and risk of fractures. On the other hand, many studies have shown a relation between high milk intake and risk of fatal or metastatic prostate cancer.[7]

The last part of the quote is why Denis decided to cut back on his milk intake.

In 1974 the Food and Drug Administration forced the milk industry to stop saying that everybody needed milk. The hugely successful (and legally unassailable) "Got Milk?" ad campaign was a joint enterprise of the Department of Agriculture and a nonprofit outfit called Dairy Management Inc., which is funded by mandatory fees on dairy farmers (not by taxpayers), and dedicated to boosting the sale of dairy products. The campaign was very loosely overseen by the arm of the USDA that promotes the consumption of dairy products.[8]

We were interested to learn that, ounce for ounce, whole milk has more calories than soda pop *and that over half its calories come from sugar.* Even a cup of nonfat milk has only seven fewer calories than a cup of Classic Coke. Chocolate milk and strawberry-flavored milk have two to three times the sugar content of plain milk.

New research suggests that heavy sugar consumption might be linked directly to cardiovascular disease and diabetes as well as to obesity. A cardiovascular disease study that was prospective, covered twenty years, and involved over 31,000 adults found that sugar added to foods (such as to dairy desserts) significantly increased participants' risk of dying of heart disease. When a participant got over 25 percent of his or her caloric intake from sugar, he or she was nearly three times more likely to die from heart disease than a participant who got under 10 percent of his or her calories from added sugar. (Most adults get over 10 percent of their caloric intake from added sugar.) Heart disease is what kills most Americans.[9]

A sugar-and-diabetes study found that the more sugar consumed in each of 175 countries, the higher the nation's rate of type 2 diabetes. More than a 1 percent increase in diabetes prevalence was found for every daily "can-of-soda-equivalent's" worth of sugar. These findings were independent of sugar's role in causing obesity and independent of couch-potato behavior or alcohol consumption. The study did not prove causation, however, and did not consider the effects of different types of sugars, such as those found in dairy.[10]

The heavy consumption of saturated fat is also associated with weight gain.[11] Over half the fat in whole milk is saturated, and we humans don't need any saturated fat in our diets. Whole milk may prompt our livers to churn out more cholesterol. Both saturated fat and cholesterol have been linked to heart disease (but this is disputed in the case of dairy, for reasons outlined below).

Most of the saturated fat Americans consume comes in the form of cheese and pizza.[12] About a third of the sixty million gallons of milk our nine million dairy cows squirt out each day is drunk as milk, and before it goes to market most of that milk has some or all of the fat removed. That means there are vats and vats of leftover fat. To find profitable uses for the fat, the dairy industry pressures other industries—such as pizza makers and burger chains—to use more cheese. The campaign has been wildly successful: According to the USDA, even as the per capita consumption of fluid milk fell, per capita cheese consumption soared from just over 16 pounds a year in 1970 to 33.5 pounds in 2012. Another huge market for dairy fat is gourmet ice cream, the segment of the ice cream market now enjoying the greatest success. A 2012 survey found that premium (high-fat, low aeration) ice cream was the industry's most popular product.[13]

The American Heart Association recommends two to three daily servings of fat-free or low-fat dairy products for most adults, and four servings for older adults and teenagers. For children, the recommendation is two or more servings of fat-free or low-fat dairy products. The USDA (MyPlate) and the American Academy of Pediatrics recommend three glasses of fat-free or reduced-fat milk daily for children. Their concern is the high rate of obesity among children. But a new study in the medical journal *Pediatrics* by Walter Willet and David Ludwig, a pediatrician at Boston Children's Hospital, questions these recommendations because of the high sugar content of milk. They suggest zero to three glasses might be a better recommendation, depending on individual needs. (Three glasses of milk contain 36.9 grams of sugar; the World Health Organization's 2014 draft guidelines recommendation for children is 12 grams a day—so even one cup puts a child over the sugar limit, regardless of whether the milk is whole or reduced fat.) The authors point out that people drinking skim milk won't feel satiated as quickly as they would if they drank whole milk, so they might eat an extra cookie or two. A diet with somewhat more saturated fat may increase energy and decrease hunger, the authors postulate, because fat isn't digested as quickly as sugar, so one doesn't feel hungry as soon. They further argue that when the goal is to reduce cardiovascular disease, replacing the saturated fat calories in milk with high-glycemic-index carbohydrate calories will increase blood levels of triglycerides, while full-fat milk doesn't change the ratio of high-density lipoprotein to total cholesterol.[14] (It would be good to replace saturated fat with unsaturated fat or low-glycemic carbs, but this isn't the American diet.)

Other researchers have found that, for unknown reasons, children's iron levels fall as their levels of vitamin D from milk intake

"EAT ICE CREAM for daily happiness," message on old Wisconsin barn. Happiness might or might not equate with healthiness. There's still a lot to be learned about dairy products. Photo: Liz Salim.

rise. They recommend two servings of milk a day for kids two to five years of age to keep these nutrients in balance. Dark-skinned kids may need up to double that because their darker skin doesn't absorb as much ultraviolet sunlight. Sunlight activates a kind of pre-vitamin D in the second layer of skin.[15]

Although saturated fat is generally bad for us, here's a baffling fact: No link has yet been found between heart attacks and *dairy* foods high in saturated fat. Milk and cheese are extremely complex substances that scientists are a long way from fully understanding. For example, micronutrients like calcium, potassium, phosphorus, and the vitamin D that is added to milk might lower blood pressure, which might reduce the probability of a heart attack. Or maybe conjugated linoleic acid (CLA) somehow protects hearts and/or mitigates some other adverse health effects of high-fat dairy food. CLA, found in whole and low-fat milk, is just beginning to be understood. Higher amounts are found in grass-fed cows than in conventionally raised animals.[16]

Three researchers at the Harvard School of Public Heath (Liesbeth A. Smit, Ana Baylin, and Hannia Campos) became intrigued by the CLA mystery. They examined fat tissue taken from 3,623 Costa Rican adults and found that those with the highest concentrations of CLA were 36 percent less likely to have heart attacks than those with the lowest concentrations. The study controlled for smoking, family histories of heart disease, and alcohol consumption. In Costa Rica, cows usually eat grass, not grain, so their milk can contain five times as much CLA as conventional American milk. The scientists concluded: "9c,11*t*-CLA, which is present in meaningful amounts in the milk of pasture-raised cows, might offset the adverse effect of the saturated fat content of dairy products."[17]

In lab animals and lab tests on human tissue, CLA inhibits tumors,[18] seems to discourage atherosclerosis and diabetes, improve the workings of the immune system, reduce inflammation, and make mice bellies firmer.[19] (Listen up, Mickey!) Some researchers think CLA's cancer and heart-disease-fighting properties stem from the way it suppresses inflammation.[20] Thus far, however, researchers have only been able to document a couple of these wondrous effects in humans. This may be because humans react differently to CLA than do rodents and rabbits, and/or because of the particular isomers of CLA used in the studies, or other of the many possible variables. (Isomers are chemicals that have the same molecular formula but different structures—CLA has at least twenty-eight isomers.) The various isomers of CLA all behave somewhat differently and may react differently in different people. For example, a study found that one isomer of CLA was effective in raising the level of "good" cholesterol, while another lowered it.[21]

CLA might conceivably help some people attain a slightly leaner body mass.[22] This theory is very controversial. For a while, Dairy Management ran adds saying that people who ate three servings of dairy products a day would lose "significantly more weight and body fat than those who just cut calories."[23] The claim was based on research by Michael Zemel, a professor of nutrition and medicine at the University of Tennessee, in work sponsored by Dairy Management.[24] If you eat a lot of dairy, you can consume more calories without gaining weight, Zemel claimed . . . *if*: (1) you burn up more calories than you swallow and (2) are not already getting enough calcium. His study involved only forty-six people.[25] Nevertheless, Dannon Light & Fit nonfat yogurt ("Slim down with yogurt") and Kraft cheese ("Burn more fat") based ads on Zemel's work, with the caution about calorie reduction shrunk to fine print.[26] Some studies support Zemel's conclusions, others don't. CLAs are tricky.

Mario Kratz, Ton Baars, and Stephan Guyenet, researchers at the Fred Hutchinson Cancer Research Center in Seattle, published a review of observational studies on the relationship between consuming high-fat dairy products and obesity, cardiovascular, and metabolic disease. They found that in eleven of sixteen studies, people who consumed high-fat dairy products were less likely to be obese and less or no more likely to have metabolic problems than people who didn't. Their findings (not conclusive) were that dairy fat or high-fat dairy foods consumed "within typical dietary patterns" are inversely associated with obesity and seem not to contribute to cardiometabolic risk. The Hutchinson researchers, who say they have no conflicts of interest, speculate that dairy fat may contain some "bioactive properties" that protect health. They warn that the studies they reviewed do not prove cause and effect and are only observational. Furthermore, the findings would not apply if you add dairy fats to your existing diet (as opposed to replacing other fats with dairy fat).

"In the meantime, it's clear that typical dietary recommendations to favor low-fat dairy over high-fat dairy are on thin ice," writes coauthor Stephan Guyenet on his blog.[27]

Before you indulge in gourmet ice cream, however, read in the next chapter about saturated fat's possible effect on your brain.

Lactose Intolerance and Milk Allergies

Some critics of milk go way beyond the problems with saturated fat and high sugar content. One is the late Frank Oski, a man with impressive credentials: former director of pediatrics at Johns Hopkins University School of Medicine and physician-in-chief at that institution's Children's Center. Dr. Oski called cow milk "the world's most overrated nutrient." He wrote: "[D]rinking cow's milk has been linked to iron-deficiency anemia in infants and children; it has been named as the cause of cramps and diarrhea in much of the world's population, and the cause of multiple forms of allergy as well . . . in no mammalian species, except for the human (and the domestic cat), is milk consumption continued after the weaning period. . . . Cow milk is for calves."[28]

Lactose is a sugar found in cow and human milk. To break lactose down into digestible glucose and galactose, our bodies produce an enzyme called lactase. Most babies make enough lactase to do this until they are at least a few years old. A white person whose ancestors lived in northwestern Europe may retain a lifelong ability to digest lactose. Adults of most other lineages, however, are likely to experience various degrees of discomfort after drinking milk: gas, diarrhea, bloating, and nausea. Milk is less nutritious to an intolerant individual because protein is lost through diarrhea, and the calories in undigested sugars are never harnessed to provide energy.

It is thought that thirty to fifty million adult Americans have some lactose intolerance. (It is probably more precise to speak of "lactose *persistence*" as the aberration, because intolerance is the norm worldwide.) Roughly 90 percent of Asian Americans and 75 percent of African Americans, Hispanics, Native Americans, and Ashkenazi Jews are lactose intolerant. Among white northern Europeans, intolerance runs only around 2 to 7 percent. (Estimates vary widely, partly because of a failure to agree on how to define lactose intolerance.) Even lactose intolerant Americans have grown accustomed to milk on cereal and cream in coffee. They continue to consume milk products because of habit, family traditions, or simply because dairy products taste good. An African American friend told us it would be easier for her to give up meat than dairy. Putting cream in her coffee is a must; she just suffers the consequences.

An allergy is an immune system malfunction, a very different problem than an inability to digest lactose. Certain proteins in milk can trigger allergies. Cow milk is the most common food allergy among children and afflicts about 2.5 percent of American children less than three years of age. By age three, about 85–90 percent of once-allergic children will have outgrown their allergy.[29] An allergy may be only to the part of milk that curdles (the casein) or to the liquid left over (the whey); or it may be to both curds and whey. Various forms of curds and whey lurk in many processed foods. A breast-feeding mother who consumes them can pass the triggering proteins on to her infant. Allergens in milk can survive pasteurization, boiling, evaporation, and ultra-high-temperature processing. Hydrolysis and further processing are needed to get truly "allergen-free" products.

Allergies can manifest themselves within minutes of an individual's consuming cow milk—in the form of wheezing, digestive problems, vomiting, hives, an itchy skin rash around the mouth, and a runny nose and eyes. Infrequently, milk allergies can even cause deadly anaphylaxis. In babies, a cow-milk allergy can cause colic. There is a positive association between allergies to cow milk and ear, nose, and throat problems in young children,[30] and a link to asthma problems later in childhood.[31]

The Calcium Paradox

People living in countries where adults seldom consume dairy products have lower rates of osteoporosis than Americans. What's going on? Why do Chinese, Africans, and South Americans have stronger bones,[32] while we in the United States consume three times as much calcium, yet have more brittle bones?

While getting enough calcium is indeed important to bones, there are factors other than diet that matter much more: skin color (white and Asian women are at greatest risk); exercise (more is better); gender (women get more osteoporosis); longevity (older people have weaker bones); being small and thin; using cigarettes, alcohol, and prednisone (all weaken bones); genetics (if osteoporosis runs in your family); and whether you have rheumatoid disease. And here are some other suspected bone-weakening factors: lack of sleep, lack of exposure to sunlight, regular sunscreen use, not getting enough vitamin K or protein, taking too much vitamin A supplement in the form of retinol (instead of the beta-carotene form), taking some common prescription drugs, or having rheumatoid arthritis. Both obesity and losing a lot of weight can weaken bones.

Still, an adequate amount of calcium is indeed important to bone health (recommended daily allowances vary by nation and authority). You can find the National Institutes of Health recommendations for Americans, set by the Food and Nutrition Board of the National Academies, on their website.[33]

Calcium intake is of special interest to postmenopausal women. Estrogen helps bones absorb calcium, and estrogen levels decline rapidly at menopause. Even women who consume a lot of calcium usually lose bone. Experts estimate half of American women will suffer a bone fracture after age fifty. Coauthor Gail is one of them, even though she took high doses of calcium/vitamin D supplement for years. (She has since greatly reduced her calcium pill intake.) Two million aging American men also have osteoporosis.

Pastured cows and big-boned elephants get their calcium from plants that extract it from the soil. So can humans. The staid Academy of Nutrition and Dietetics says vegan humans can get adequate calcium from low-oxalate greens like bok choy, broccoli, Chinese cabbage, kale, and collard greens, and from calcium-set tofu.[34] (True, milk is a convenient source—one cup of milk contains as much calcium as two and one-third cups of kale.)

Cooking greens is a bother. So if you don't want to consume dairy foods, why not just pop calcium pills? Because supplemental shortcuts don't seem to work as well as getting calcium from food. Large studies have found no evidence that taking calcium and vitamin D supplements significantly reduces bone loss or hip fractures.[35] In middle-aged and older women, a recent meta-analysis also found that taking calcium supplements might increase the risk of having a stroke.[36] (Kidney stones are another risk of a high calcium and/or oxalate intake.) If you're already getting around 1,000 mg per day of calcium through your food, there's no need to take calcium supplements.

In 2013, the U.S. Preventive Services Task Force surprised many people when it announced its findings after a review of over one hundred studies: When post-menopausal women take a daily pill containing 1,000 milligrams of calcium and 400 international units of vitamin D, they did not experience fewer fractures but did increase their chances of getting kidney stones. Another consideration is a 2012 long-term study involving twenty-four thousand German men and women, which indicates that getting calcium from pills might increase the risk of a heart attack.[37] Therefore, the Task Force "recommends against daily supplementation with 400 IU or less of vitamin D_3 and 1,000 mg or less of calcium for the primary prevention of fractures in noninstitutionalized postmenopausal women." The findings did not consider women who already have osteoporosis, those who are deficient in vitamin D, or those who took higher doses of the supplements.[38] Not all experts agree with the U.S. Preventive Services Task Force's recommendations, but there does seem to be widespread agreement that it's better to get calcium from foods than from pills. Aim for getting about half your calcium from real food.

Germs and Chemicals in Milk

Consider where the udder is located on a cow and what else is nearby. Although teats are cleansed prior to milking, mistakes happen. Or a sick cow might already have germs in her milk. Or someone processing milk might get careless. This is why milk is pasteurized.

The Food Safety Modernization Act was signed into law in 2011 and finally gave the FDA authority to swiftly recall many contaminated foods, including dairy products. The FDA has been slow in issuing regulations under the law. In 2013, the Office of Management and Budget (OMB) extracted most teeth from the new law, which still hadn't gone into effect. The OMB removed the requirement for verified food safety plans and made testing for contamination in foods and at processing facilities voluntary. This action was done so stealthily that not even an announcement was made. A whistleblower in the Department of Health and Human Services (which oversees the FDA) posted the OMB's changes on the Internet.[39]

When bulk tank samples of raw milk were tested in recent years, pathogenic organisms were found in 0.87 percent to 12.6 percent of samples.[40] Early in the twentieth century, the diseases people commonly got from consuming raw milk products were brucellosis, diphtheria, typhoid, and tuberculosis. Today the danger is more about salmonella, listeriosis, and diseases caused by *Campylobacter jejuni* and *E. coli* O157.

Raw milk lovers are passionate about their right to purchase and drink unpasteurized milk, which they believe to be healthier than pasteurized. Whether it tastes better is a matter of individual preference. We had an opportunity to drink some fresh, chilled raw milk at a small dairy we trust and found it so good we each guzzled an entire twenty-four-ounce glass. To us it tasted much better than the low-fat, pasteurized milk we buy. It was richer and—like gourmet ice cream—left a creamy aftertaste.

Discussions of raw milk often lead to raw feelings among friends. Advocates of raw milk argue that drinking raw milk usually supports local, smaller dairies that are more likely to be held accountable to their customers and to have healthier, happier cows and better farming practices.

But you assume some risk when you drink raw milk. According to the CDC, from 1998 through 2009, consuming raw milk or raw-milk products caused 93 outbreaks of disease resulting in 1,837 illnesses, 195 hospitalizations, and 2 deaths.[41] Surveys like this have persuaded the USDA to ban raw milk from interstate commerce. State law, however, governs sales within a state, and selling raw milk is legal under differing circumstances in over half the states.

Many advocates argue that raw milk is more nutritious. Most scientific studies, however, indicate that few important nutrients are lost during pasteurization. A meta-analysis of research papers on the nutritive value of raw vs. pasteurized milk concluded that pasteurization does minimal harm to vitamins. Pasteurization decreased the level or quality of vitamins B_{12} and E. It increased levels of vitamin A. The levels of vitamins B_1, B_2, B_9, and C were somewhat decreased. Milk isn't an important source of vitamins B_9, B_{12}, C, or E, but is an important dietary source of vitamin B_2. Six of the studies found that children who drink raw milk may have less asthma, hay fever, and eczema (but it's hard to separate drinking raw milk from other farm-life factors). No association with cancer was found. The poor quality of many of the forty studies included, however, made these findings less convincing than they would otherwise have been.[42]

Fans of raw milk commissioned a study by Stanford University researchers that they hoped would prove raw milk caused fewer problems in lactose intolerant people than pasteurized milk. But the study found no significant differences in symptoms experienced by people who consumed the two types of milk.[43] There may be other health benefits of raw milk, however, that haven't been studied.

The wisdom of drinking raw milk appears to be one of those issues on which most people seldom change their opinion. But children aren't capable of making an informed decision, and a milk-borne infection can be deadly. In December 2013 the American Academy of Pediatrics issued a policy statement urging pregnant women, infants, and children not to drink raw milk or consume raw milk products. The Cornell University Food Science Department has issued a similar position statement: "[W]e strongly recommend that raw milk not be served to infants, toddlers, or pregnant women, or any person suffering from a chronic disease or suppressed immune system."

The Cornell researchers also oppose the sale of raw milk to the public.[44] We disagree. The risk of death for adults presently appears to be low, there is still much to be learned about the possible benefits of raw milk, and adults might want to assume that risk. As for ourselves, however, we buy pasteurized milk. Having had a frightening encounter with antibiotic-resistant bacteria when Denis was infected with MRSA (chapter 4), we're aware that in the near future some infections may be impossible to treat with antibiotics.

Cheeses made from raw milk can also contain harmful bacteria. Contamination can come from the milk or occur during cheese manufacture, storage, and packaging. But the consumption of raw cheese presents more nuanced issues than does the consumption of raw milk. Many artisan cheesemakers say good bacteria and enzymes that add flavor to their cheeses are destroyed by pasteurization. Aging some types of cheese may kill harmful bacteria. The Code of Federal Regulations sets detailed standards for over ninety types of cheese.[45] Under this law, soft-ripened cheeses, semi-soft cheeses, firm/hard cheeses, blue cheeses, and natural or washed rind cheeses made from unpasteurized milk, but aged at least sixty days, are legal. (Sixty days gives acids and salts in cheese time to destroy bacteria, and as cheese dries it becomes less hospitable to bacteria.) But the sixty-day rule isn't based on hard science, and the FDA is reconsidering it. Bacteria that make people sick have been found in cheese aged sixty days or longer.[46]

In 2010, thirty-eight people fell ill after eating raw-milk Gouda cheese bought at Costco that was contaminated with toxic *E. coli.* Reportedly, Bravo packaged its Gouda before it had aged sixty days.[47] In another outbreak, toxic *E. coli* in raw-milk-based soft cheeses from Oregon made eight people sick. Unsanitary conditions might have been involved in the latter case.[48] David Gumpert, a guest contributor to *Grist*, noted that when you eliminate outbreaks caused by queso fresco (an illegal soft cheese that isn't aged), the difference between outbreaks associated with raw-milk cheese and those associated with pasteurized-milk cheese seem less dramatic. From 2000 to 2009 there were 350 illnesses and no deaths associated with raw-milk cheese outbreaks, and 247 illnesses and one death associated with pasteurized-milk cheese. (This doesn't consider that more pasteurized-milk cheese is consumed.) Two authorities are quoted in the article as saying most problems are caused by lack of sanitation during manufacture and post-production contamination. The artisanal cheese industry has been growing rapidly and some producers may still be in the learning stage.[49]

A few types of bacteria can survive pasteurization. One is *Mycobacterium avium subspecies paratuberculosis* (MAP), which causes Johne's disease in cows. It might also be connected with Crohn's disease in humans, although this theory is controversial. Crohn's causes inflammation of the gastrointestinal tract and torments over half a million Americans with diarrhea, abdominal pain, and vomiting. There is no cure. A genetic predisposition, an infectious agent, and exposure to things in the environment are all implicated in Crohn's. Curiously, this disease only showed up in the twentieth century, and the incidence is rising.[50]

MAP hides inside white blood cells and thus is somewhat protected from the heat of pasteurization, even ultra-pasteurization.[51] In 2007, the USDA's Animal and Plant Health Inspection Service found that in nearly a quarter of dairy operations surveyed, at least 10 percent of the cows were infected with Johne's. Compared to operations of five hundred or more cows, smaller operations with fewer than a hundred cows were half as likely to have infected cows in their herds.[52] MAP is excreted in cow poop, can survive for a year in the environment, and from there it may get into milk, groundwater, tap water, and food crops.[53] Studies have shown it can infect humans.

Scientists on both sides of this controversy present strong cases. A study of forty Crohn's patients and forty persons without the disease found that the Crohn's patients had many more MAP bacteria in their guts than the control group.[54] Clusters of Crohn's disease patients have been correlated with MAP in water supplies (chlorination doesn't kill MAP).[55] On the other hand, when tissue from Crohn's patients is compared to tissue from cows with Johne's, differences are found. And people who work with cows don't seem any more likely to get Crohn's than other people. The MAP organisms found in Crohn's patients may simply be bystanders in a sick bowel. A good review of this issue can be found online at the University of Wisconsin Johne's Information Center.[56]

If you're of an apocalyptic bent, here's a scenario that will grab your attention. Two Stanford University researchers, Lawrence M. Wein and Yifan Liu, wrote an article for the *Proceedings of the National Academy of Sciences* contending that milk is an ideal transmitting agent for biological terrorism.[57] (Wein summarized their findings in a *New York Times* op-ed.)[58] The weapon: botulinum neurotoxin, the most toxic biological substance known. It blocks neuromuscular transmission and is appallingly easy to obtain and prepare for distribution.[59] One of its trade names is Botox. Many Americans get injections of Botox to reduce facial wrinkles.

Every year, six billion gallons of milk are pooled at various stages of collection and shipped around the country for rapid consumption. A terrorist could pour a few grams of botulinum toxin into a milk tank on one or more farms. The contaminated milk would be taken to a factory and poured into a raw-milk silo along with milk from other farms. From there it would flow into an even larger product stream.

One-millionth of a gram of botulinum toxin can kill an adult; even less suffices to kill a child. This scenario holds the possibility of killing hundreds of thousands of people, Wein and Liu believe, most of them children. They argue that all milk trucks should be tested for botulinum before they unload into a silo, noting that trucks have to stop anyway for tests of antibiotic residue. Testing for all four types of the toxin that harm humans would raise the cost of a gallon of milk by less than one cent.

Wein and Liu's article, which appeared in 2005—after the anthrax attacks of 2001—caused quite a stir. Some people complained that they were providing advice to terrorists. In the same issue of the *Proceedings* in which the article was published, Bruce Alberts, then president of the National Academy of Sciences, noted in an editorial that the information in the article was already freely available on the Web. He suggested that public discussion would help create greater alertness to an existing threat and acknowledged that there had been improvements in bioterrorism safeguards since 9/11. Furthermore, Alberts said, Wein and Liu may have underestimated how much of the botulism would be destroyed by newer pasteurization standards, a fact pointed out by other scientists.[60] Alberts also noted, however, that these improvements in pasteurization, as well as other important FDA guidelines for protecting the milk supply, are voluntary. Not all milk producers have adopted them.[61]

Many people believe that such threats to our milk supply are remote, and that protecting against them is too expensive—another example of an overreaching nanny state. Others, including the authors, note that milk terrorism requires much less money and a far lower level of coordination, intelligence, and discipline than training jetliner pilots, simultaneously commandeering four transcontinental jetliners, and crashing those airplanes into specific buildings. If we can prevent the possible killing of thousands of people at a reasonable cost, let's do so.

There is also a naturally occurring toxic substance in milk that's now mostly of historical interest—tremetol. This poison is found in the white snakeroot plant. When eaten by cows, it gets into their milk, and when humans drink that milk it causes trembling, horrid intestinal pain, and vomiting. Abraham Lincoln's mother and thousands of other settlers died of this "milk fever." Although tremetol is not inactivated by pasteurization, the milk in a single carton today comes from so many different cows that any tremetol would be too diluted to be dangerous. The risk of milk fever today is exceedingly slight.

Hormones are another invisible substance in milk. They are found in all milk, be it bovine or human. Our bodies, and cows' bodies, are regulated by an intricate symphony of hormones essential to our growth, sexual maturation, and health. A master gland, the pituitary, conducts the hormone production of other glands. What we ingest can throw this system out of tune.

Toxicologists' mantra has long been, "The dose makes the poison." Ordinarily, the higher the dose, the greater the response—hence all those tests stuffing lab rats with huge amounts of food additives. But this axiom is not true of hormones, nor of chemicals that interfere with, or mimic, hormones. Scientists cannot predict the effect of low doses of hormones from effects seen when a high dose is given.

Estrogen is one key hormone. All animals with backbones make estrogen. In women, estrogen regulates estrus cycles and the development of breasts and other secondary sexual characteristics. In males, it's important for the maturation of sperm. Without estrogen, neither cows nor human females could make milk. A milk cow is usually "freshened" (bred) about once a year. Unlike Grandma's cow, she keeps lactating in the last half of her pregnancy. Cows produce about thirty-three times more natural estrogen when they are pregnant than they do when they are not pregnant, which means there's a great deal of estrogen in the cow milk you drink.[62] As a cow's pregnancy advances, the estrogen content of her milk rises. Americans drink so much milk that it accounts for 60–80 percent of the estrogen and progesterone (another female sex hormone) that we take into our bodies. Most of the rest comes from other animal products. (Plant phytoestrogens sometimes act like estrogen and sometimes like an antagonist to it.)

In men, too much estrogen can mess with male reproductive systems and cause breasts to grow. Sperm banks in Israel are finding it harder to get healthy sperm; only 1 percent of would-be donors at that country's largest sperm bank make the grade. The cause of this alarming decline in fertility is a mystery. It's thought by some scientists that aggressive dairy farming methods lie behind the drop.[63]

Excessive estrogen might mean an increased risk of ovarian cancer in women, prostate and testicular cancer in men, and breast cancer in both. In countries where fewer dairy products are consumed, young men have much less testicular cancer.[64] The Harvard School of Public Health advises: "Because of unresolved concerns about the risk of ovarian and prostate cancer, it may be prudent to avoid higher intakes of dairy products."[65] Hormones reside in fat; if you choose to lower your level of estrogen exposure, pick low-fat or nonfat dairy products. Coauthor Denis has done so.

In the 1960s, a synthetic estrogen, diethylstilbestrol (DES), was given to pregnant women to prevent miscarriages and premature deliveries. It was also given to beef cows to promote growth. DES had to be withdrawn from the market because it greatly increased the probability that a daughter of a woman who took the drug would develop vaginal and cervical cancer, be a little more likely to have breast cancer, be born with odd reproductive-tract structures, and experience pregnancy complications or infertility. Although most attention was paid to the serious problems of the so-called DES daughters, DES boy babies had their own, lesser problem: They were more prone to noncancerous cysts in the tubes that carry sperm from the testicles. The DES tragedy heightened public awareness of the sucker punches hormones can throw. Therefore, many people were upset when American dairy farmers began treating their cows with recombinant (synthetic) bovine growth hormone in 1994.

Bovine somatotropin regulates metabolic processes in cows. It is not recognized by hormone receptors in humans, even if injected, and when consumed as food it does not seem to survive the digestion process.[66] The synthetic versions of this hormone are called recombinant bovine somatotropin (rBST) or recombinant bovine growth hormone (rBGH). Some versions might differ from natural BST by one amino acid.[67] Cows treated with rBST are said to produce 11 to 16 percent more milk than those who are not. The dairy industry says that pasteurization destroys 90 percent of both natural and manufactured bovine growth hormone.[68]

Most dairy cows are not treated with rBST, however, because consumer-sensitive retailers such as Costco, Ben & Jerry's, Kroger, Safeway, Starbucks, and Walmart refuse to buy such milk. Most current estimates are that 17 to 18 percent of dairy cows are injected with rBST (usually by a jab in the tail region). But because milk is pooled, if it isn't labeled "no rBST" it's hard to know what is inside a container. Organic dairy products do not contain rBST.

The dairy lobby leaned on states to make it illegal to label milk "rBGH-Free" or "rBST-Free." But in the fall of 2010, the Sixth Circuit

Court of Appeals held that Ohio could no longer ban such labels. The court also found that rBST-produced milk is different in composition from regular milk because it contains more pus, spoils more quickly, and is lower in nutritional quality at some points in the cow's lactation cycle.[69]

Concerns about the effect of rBST on the health of both humans and cows have led the European Union, Japan, Australia, and Canada to ban its use. But the United States' National Institutes of Health, the FDA, and the EPA all maintain that milk produced with growth hormones is safe for human consumption.

The main concern about the impact of rBST use on humans revolves around another hormone that rBST causes cows to produce more of: insulin-like growth factor 1 (IGF-1). IGF-1 is produced naturally, in an identical form, in both humans and cows. It is also used by some professional body builders to beef up, and is administered to children with rare diseases, such as Laron dwarfism. Having just the right amount of IGF-1 is particularly important for babies, because IGF-1 helps to control the rate at which cells divide and differentiate. The FDA declares it safe to give milk from rBST-treated cows to infants and children.[70] But the revolving door between the drug industry and the FDA may have compromised the agency's objectivity.[71] Some American scientists join public health agencies abroad in questioning the safety of giving growing children milk from rBST cows.[72] A major, well-designed study, involving researchers from universities and institutions in Great Britain, Europe, and the United States and 4,731 men and women who agreed to be studied, found that the level of IGF-1 in a person's blood rises along with a person's consumption of dairy products.[73]

In addition to dairy consumption, a great many things affect IGF-1 levels in humans: gender, genes, age, the time of testing, exercise, stress, caloric and animal protein intake, body mass index, and on and on. The levels of IGF-1 in cows vary similarly. The natural variation in cows is greater than the increase in IGF-1 caused by giving cows rBST.

Research is less ambiguous when it comes to the impact of rBST on cows: It is bad for them. Monsanto's own packaging lists twenty possible side effects of rBST on cows, including reduced hemoglobin and skin rashes. A report from the Canadian Veterinary Medical Association says rBST increases lameness (especially in the leg joints) by 50 percent and clinical mastitis (udder infection) by 25 percent. An udder infection makes it more likely a cow will have pus in her milk and need antibiotics. Cows treated with rBST were more likely to be slaughtered at a younger age, especially if they had had more than one calf.[74]

The bad effects of rBST on cows are the main reason its use has been banned in most of the industrialized world. It would be a kindness to cows to ban its use in the United States as well.

The hormones produced by animals' bodies have Frankenstein first cousins: manufactured chemicals that act like, or mess up the action of, natural hormones. Called "endocrine disruptors," they can be found in plastics, cosmetics, toys, pesticides, herbicides, linings of metal food cans, flame retardants, and many other products. They escape from products, get into air and water, and hence into food and our bodies. Even human breast milk now contains endocrine disruptors. Women who breast-feed are generally horrified to learn that they will lower their own load of some endocrine disruptors by transferring them to their babies.[75] (Breast-feeding is still strongly recommended because of its many other benefits.) Because the effects of low doses of hormones and endocrine disruptors can't be predicted from studying high doses (as is done with other cancer-causing chemicals), the whole approach to evaluating their safety has to be different. High doses of them might occasionally be harmless, but tiny doses cause lasting effects![76]

An endocrine disruptor can act alone or synergistically, in combination with other endocrine disruptors. They can have a big effect in minuscule amounts—often not noticeable until many years after exposure. This makes them devilishly hard to understand.

How do endocrine disruptors get into cows? Most of them probably come from cow feed. They accumulate in cows' fat.[77] Pesticides that are known to be or are suspected of being endocrine disruptors, and which are applied to crops used to feed cows, include the five leading herbicides applied to corn and soy in the United States: atrazine; glyphosate; 2,4-D; acetochlor; and metolachlor.[78] (Organic farmers aren't allowed to use these chemicals, so the risk is almost eliminated in organic dairy and beef products.)

Another endocrine disruptor is perchlorate. Perchlorate is very harmful to thyroid glands, which play a key role in growing brains. In the spring of 2009, the CDC found perchlorate in all the major brands of infant formula sold in the United States. Perchlorate is in rocket fuel, and it gets into the environment during rocket and missile tests. Perchlorates are also used in the carpet industry, explosives, and a variety of industrial processes. Perchlorate has contaminated the water supply in over half the states and thus gotten into cows and human babies. Tainted infant formula powder mixed with tainted water, the CDC warns, could constitute an unsafe dose for a baby.[79] A study of female rats found that when they were exposed to the levels of perchlorate found in our environment, the rats' thyroid glands and reproductive systems were affected.[80]

Dioxins are yet another type of endocrine disruptor. They occur naturally but are also released by fires and industrial activities. Dioxins can cause cancer, birth defects, genetic mutations, and thyroid, reproductive, and immune system disorders. According to the World Health Organization, over 90 percent of humans' dioxin exposure comes from nonorganic meat, fish and shellfish, and dairy foods.[81]

In their groundbreaking *Our Stolen Future* (1997), Theo Colborn, Dianne Dumanoski, and Pete Myers presented an overview of some probable side effects of this chemical storm of endocrine disruptors: a dramatic drop in male sperm counts in parts of the world over the last two generations, defective sexual organs, lower IQs, behavioral abnormalities in children, more cancers, and gender-ambiguous wildlife.

An individual's exposure to endocrine disruptors at critical times can affect multiple generations, probably because of epigenetic changes (particular genes being turned on or off). To be clear, *something you eat or drink before conceiving a baby or while pregnant or nursing might, many years later, affect the health of your great-grandchild.* Think about that. This warning applies to future fathers as well as to mothers.[82]

Endocrine disruptors can sometimes cause delayed reproductive maturation in boys.[83] In girls, some scientists think endocrine disruptors help trigger the too-early development of breasts and pubic hair and the onset of menstrual periods. American girls today get their periods, on average, several months to a year earlier than they did forty years ago, and they develop breasts a year or two earlier. At age eight, roughly 18 percent of white, 43 percent of black non-Hispanic, and 31 percent of Hispanic girls in the United States have entered puberty.[84]

It's theoretically possible to have "milk" without cows. A small start-up outfit with this goal calls itself "Muufri" (get it?). Milk brewed like high-tech beer wouldn't merely avoid waste lagoons, corn production, and the potential for animal cruelty—the end product would also be free of pesticides, antibiotics, synthetic hormones, and the other problems highlighted in this chapter. It could even be designed to eliminate lactose, cholesterol, and all bacteria.

But is this realistic? The fact that no one has done it yet suggests that it isn't easy. Still, compared to developing cow-free meat, it appears relatively straightforward. There are just four simple proteins in the casein component of milk. They are small and reportedly easy to make. Whey has two principal proteins that are only somewhat more complex to engineer. If you introduce the DNA for those proteins into yeast, the yeast will start producing the proteins, much like yeast does for beer. As with beer, the yeast will need a source of nutrition, but the conversion efficiency of yeast is vastly more efficient than that of cows. Mix some plant-based oils and fatty acids into the final product, along with a dash of sugar and trace ingredients for taste, and you're good to go. Of course, many things that look good on paper and produce product in the lab run into problems when scaled up. But Muufri got a 2014 accelerator grant from SynBio axlr8r, and is off and running.

If marketed simply as a commercial liquid rich in protein, low in fat, and lactose-free, synthetic "milk" might be a welcome addition. It would doubtless be better for many people and better for the environment than sodas, energy drinks, frappuccinos, and most other beverages. But it shouldn't be called "milk." As discussed, real milk is an exceedingly complex substance.[85]

Say Cheese

Dairy farmers today are, for the most part, fully informed on all the issues about milk we've raised here. Some change their practices; some continue in the milky mainstream because they believe the risks are overstated and small in comparison to the benefits of dairy foods in American diets. Others would like to perfect their product and treat their cows better but can't afford to.

Those visionaries who plunge in and try to do things right often have steep learning curves. Kurt Timmermeister, for example, had no background in farming when he sold his Seattle restaurant and bought a few abused acres on Vashon Island, a short ferry ride away from the city. One of his first decisions as a farmer was to buy a cow, Dinah. Dinah gently taught Kurt a lot about farming and milk production. Kurt taught himself about small-farm economics. He learned the hard way.

Early in his adventure, he planted a garden on his property. "Kurtwood Farms" bought a few sheep and goats and various other animals, and tried selling milk and vegetables. Unlike a corn or soy farmer, he got no help from the federal government but a lot of hassle from local and state authorities. They particularly gave him trouble about selling raw milk and serving weekend dinners to a couple dozen loyal customers, insisting he build the bathroom facilities necessary for a much larger operation.

Kurt had to discontinue the operation of his small restaurant. He also decided that the chemistry between him and sheep and goats wasn't good, and learned how exceedingly hard it is to make a living selling veggies. But he kept Dinah. He likes cows. So he focused his efforts on making gourmet cheese. Slowly, he built up a herd of thirteen Jerseys and bought more land until he had thirteen acres. With one cow per acre there's no need for a waste lagoon on this farm.

The driveway to Kurtwood Farms is unpaved and unmarked. We drove through a small copse and when we emerged we saw two six-week-old steers gamboling around a corral. The animals were clearly playing together, having a wonderful time on this sunny spring day.

Kurt walked out to meet us by the corral. A single man with a trim, youthful air about him, Kurt has close-cropped hair and wore blue jeans and mud boots. But he also wore a long-sleeved robin's-egg-blue shirt under a well-fitted loden-green vest, which suggested greater fashion sensitivity than most farmers. As we walked around with him, the farm and beautifully restored pioneer farmhouse also reflected an awareness of aesthetics.

Our first stop was at the new cave, needed to store wheels of cheese at a cool temperature while they age. The cave is dug into a hill and has enough rebar and concrete for a bomb shelter.

"This cave will be here forever," Kurt said. Inside were tidy rows of wooden shelves covered with rounds of cheese. From there we

A handful of cows and closeness to the earth provide a good life for this small farmer. The new cheese-ripening cave at Kurtwood Farms. Photo: Gail Boyer Hayes.

walked out to a pasture to meet the juvenile steers' single moms. Above, the sky was cerulean blue, the clouds appropriately puffy and white. Underfoot, the grass was fresh and green. The mild-tempered ladies were lying down when we arrived, chewing cud. Large Jerseys with half-full udders, they didn't bother to get up and just eyed us lazily. The cows all have names and don't mind being petted. Only seven cows were then providing milk, Kurt told us.

Several cows had horns that were slightly askew, giving them a mildly disheveled appearance. We asked about the horns; it's unusual to see Jerseys that haven't been polled (had their horns removed). Kurt replied that when he was starting out someone told him cows with horns produced more milk. That lore turned out to be false. Jerseys might be smaller than Holsteins, but these lovely ladies are still strong animals who can weigh up to 1,000 pounds (bulls can grow to 1,800 pounds), and they can knock over, gore, or even kill a guy just by walking into him. When they jostle to be first in line to be milked, horns can rip hides. But removing horns from grown cows is difficult and painful, so the horns remain.

The farm isn't organic because it costs too much to buy and truck organic hay to the island during the rainy season. Antibiotics are used sparingly when cows get sick and to prevent infections when they enter their "dry" periods. (Farmers stop milking cows late in a pregnancy so a cow's energy can be diverted to growing her calf. Udders undergo drastic changes and are very vulnerable to infection during this two-month period.)

When his first attempts at living off the land didn't work, Kurt followed in the footsteps of many other small farmers and focused on a value-added product: He is now the only maker of Camembert cheese in King County, Washington. Back at the building that once housed the restaurant, he offered us a sample of Dinah's Cheese. The mouthfeel and taste were irresistible.

Although he said he isn't building up a big nest egg (he reinvests profits in the farm), neither is he going broke. The secret to his success is his considerable people and marketing skills. Kurt is friendly and outgoing, and he learned a lot as a restaurateur that now helps him. Twice a week, he loads his truck with cheese and makes the half-hour ferry trip into Seattle, the city where he sells 90 percent of his product. By building his own contacts, he avoids paying 20 percent to a distributor. Some cheese is sent to a distributor in Oregon and some shipped to California and New York.

Delivery days are Kurt's favorite time of the week. It gets lonely for a single man on a family-centered island like Vashon, and Kurt enjoys seeing his buyers and dealing with any questions on the spot. He sells to restaurants and stores, with most sales made to small chains. It takes years, he says, to develop a loyal following for a cheese. He and his employees occasionally hand out samples at stores or cheese festivals.

Food and cows matter to Kurt, but he isn't overly sentimental. The way he sees it, cows are fairly dim-witted and care only about food and sex. And although his cows live longer than typical milk-factory cows, at some point they are slaughtered on the farm and become meat.

"My hope is for people to know what a cow is," Kurt tells us, "and that cheese can be made here and not just in a factory in Wisconsin." He shows schoolchildren his milking parlor, which is just a single stanchion, a one-cow-at-a-time vacuum pump, a bucket, and a milking stool. The bucket of milk is carried a few feet to a room where it's pooled and chilled in a thirty-seven-gallon stainless tank made in Slovenia. Then it is pasteurized. From there it goes to a cheesemaking room under the same roof. For sanitary reasons, the two cheesemakers never set foot in the milking parlor or the milk-storing room.

In addition to the cheese workers, Kurt employs a man to handle the cows and pasture. The downside of being a cheesemaker, Kurt said, is the hard work. Something always needs doing. By hiring help he can make time to write books, like *Growing a Farmer.*[86]

Although he still has faith that five-, ten-, and fifteen-acre farms can thrive and have some impact, he doesn't think all small farms are good and all big farms bad. It bugs him that some writers assume that farmers are stupid, and that they don't care about taking care of their land. "They are either trapped by loans or by contracts and a Walmart culture that will pay only eighty cents for a pound of beef." (Dinah's Cheese sells for about $14 a half-pound.)

His advice for young, would-be farmers: Become a lawyer or orthodontist, make a pile of money, and *then* retire as a dairy farmer. You don't want to start out with a mortgage and with payments com-

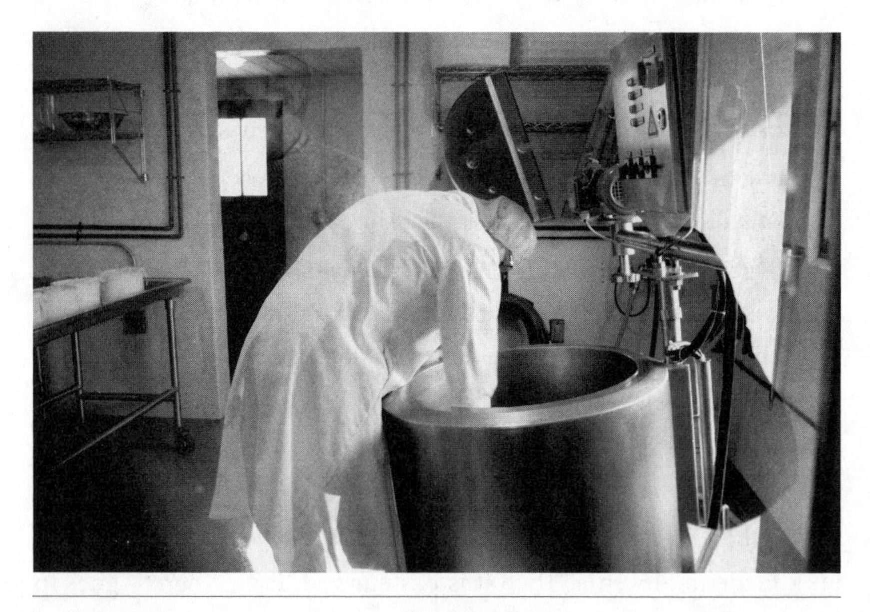

Cheesemaker at Kurtwood Farms. The ghost above the worker is a reflection in the glass separating the clean cheesemaking room from the milk-cooling room. Sanitation is critical in the dairy business. To ensure this, cheesemakers have their own entrance so they don't pass through the milking parlor or cooling room. Photo: Gail Boyer Hayes.

Kurt Timmermeister and friendly calf. Kurt tries not to be sentimental about his cows but still feels a close bond with them. Two of his cheeses are named after his cows. Photo by Gail Boyer Hayes.

ing due before you have anything to sell. And look for land near your markets.

What he likes best about farming is the outdoors life, being his own boss, and making something real that people like to eat. "I love this. I like to keep things simple," he says. "Milk the cows twice a day, make cheese, take it to stores, cash the checks."

Chapter Six

DON'T HAVE A COWBURGER

Men smell better and have more mating opportunities if they don't eat red meat—at least that was the conclusion of a 2006 Czech study. Seventeen males were put on red meat or nonmeat diets for two weeks while their armpit odors were collected on pads. Thirty biddable women then sniffed the pads and made pronouncements on how attractive, pleasant, and intense the odors were. A month later, the same men spent another two weeks on diets that excluded all meat and again had their armpit odors assessed. The researchers concluded: "[R]ed meat consumption has a negative impact on perceived body odor hedonicity."[1] In lay language, the women thought red-meat-eating guys stunk.

An alternative view can be found in Betty Fussell's *Raising Steaks*: "Real American men, women and children eat steak because it is red with blood, blood that pumps flavor, iron, vitality and sex into flaccid bodies. For women steak is better than spinach. For men, it's better than Viagra."[2]

Combine the two views and you get the worst possible outcome: Red meat arouses men but makes them repellant to women. Nevertheless, people have been eating beef for eons and somehow managing to reproduce.

If you're a beef eater and starting to paw the ground because you think we're going to tell you not to eat beef, relax. We're not anti-beef. What this chapter argues is that Americans should eat much less beef and what we eat should be of a higher quality and produced in a more humane, sustainable manner.

A popular hypothesis is that our metabolically expensive big brains wouldn't have evolved without convenient access to meats that are easily digestible and rich in nutrients.[3] We evolved to be omnivores. When humans are just sitting around, our brains (which account for only 2 percent of our weight) demand 20 percent of our bodies' energy. Unlike cows or gorillas, we don't need to spend hours and hours finding and eating plants to meet our caloric requirements. Another pro-beef argument is that in an increasingly populated world we really can't afford to let pasture land that isn't suitable for growing fruits and vegetables go to waste. And waste from vegetables—the parts humans can't or won't eat—might also go to waste if not fed to animals.

Beef is a convenient source of things our bodies need, including protein, iron, vitamin $B_{12,}$ and zinc. As Betty Fussell noted, beef is loaded with iron. A single "quarter-pounder" comes close to providing the recommended daily allowance (RDA) of iron for men and post-menopausal women.[4] But beef (or any meat) is no longer essential to superb health; adults can easily get their RDAs from beans, seafood, nuts, or other meats. An excellent selection of plant-based foods is now available year-round, and information on how to combine plant foods to get all essential nutrients is readily available. The carbon footprints of plant-based foods are far smaller than the footprints of meats. Among meats, beef and lamb, regrettably, are the champs at maxing out footprint size. (If you don't already have it, get the Environmental Working Group's *Meat Eater's Guide to Climate Change & Health*, which is available on their website.)

Although protein is essential to human health, a surprising new study found that too much protein can shorten lifespans. In 2014, a peer-reviewed, two-decade study of 6,381 adults between fifty and sixty-five years of age discovered that those whose diet was high in animal protein had the same cancer risk as cigarette smokers. Such "paleodieters" who obtained 20 percent of daily calories from a diet rich in animal protein were four times as likely to die of cancer during the study period as those who got no more than 10 percent of their calories from protein.[5] Those on a high-protein diet were also three times as likely to die of diabetes, and 74 percent more likely to die of any cause. Among the participants, *plant* protein intake neither worsened nor improved rates or cancer and mortality; nor did fat or carbohydrate intake.

Another key finding of this study was that people over sixty-five did not see reduced cancer or mortality if they ate a low-animal-protein diet, although they did experience reduced diabetes. It is thought that this older group benefited from a moderate to high amount of protein because protein helps guard against frailty. (Levels of the growth hormone IGF-I are controlled by protein, and after age sixty-five, levels of this hormone drop off dramatically in humans.)[6]

In an interview with *ScienceDaily*, USC gerontology professor Valter Longo, the study's lead author, concluded, "The majority of Americans are eating about twice as much protein as they should, and it seems that the best change would be to lower the daily intake of all proteins but especially animal-derived proteins," at least in middle age. "But don't get extreme in cutting out protein; you can go from protected to malnourished very quickly."[7]

As is true of most types of food, some beef and dairy is good for us, while too much is bad. A quarter-pounder hamburger patty made from grass-finished beef provides 216 calories and 20 grams of protein. An eight-ounce serving of nonfat milk would add 83 calories and 8.3 grams of protein. This would already exceed the amount of protein that Valter Longo's study recommends for middle-aged people, but it's not a terrible lunch (if you eat veggies and fruits for dinner). A typical teenager meal, however, looks more like a Triple Whopper with a chocolate milkshake. The nutritional difference is staggering: 2,104 calories (nearly 300 from fat) and 66 grams of protein.[8] And that doesn't include any fries. This might be a reasonable caloric intake for a sedentary person for an entire day, but most people also enjoy having breakfast and dinner. And it's too much protein for anyone.

Americans eat more cow per capita than the inhabitants of any other country except Argentina and tiny Luxembourg. Most cow meat (a whopping 42 percent) is eaten in the form of ground beef and the disinfected, mechanically separated beef nicknamed "pink slime." The rest is eaten in the form of steak (20 percent), stew (13 percent), processed forms (13 percent), and as beef dishes and other cuts (12 percent).[9]

Denis's father, Archie, dropped out of school in the sixth grade to support his own widowed mother and worked his whole life in paper mills to provide his family with a middle-class income. A second-generation Irish American, he wanted his family to eat beef for dinner six nights a week (Friday was fish night). Archie was not an early Dr. Atkins devotee—eating beef was then a symbol that you'd made it. This is no longer true. Today, the richer you are in America, the *less* beef you are likely to eat.

A 2005 study found a dramatic difference between males and females in beef consumption: Males ate, on average, 85.7 pounds of beef a year, females 48.1. Of course, men are generally larger (averaging 195 pounds) than women (165 pounds), but not enough to account for the difference in beef intake. Younger men aged twenty to thirty packed away nearly 110 pounds of beef a year. African

Americans ate more beef (77 pounds a year) than European Americans (65 pounds). Rural folks ate more (75 pounds) than urban and suburban dwellers (63 pounds). Midwesterners (73 pounds) ate more than northeasterners (63 pounds).[10]

The typical American eats far more protein than is necessary.[11] In fact, the typical American eats too much everything. Two-thirds of us are obese or overweight. The lower retail food prices brought about by "technological innovations in agricultural production" (a near synonym for "factory farms") are about 40 percent responsible for the unhealthy food choices and fattening up of Americans.[12] Obesity is associated with type 2 diabetes; coronary heart disease; hypertension; stroke; liver and gallbladder disease; sleep apnea; osteoarthritis; and colon, breast, and endometrial cancer.[13] A recent (controversial) parsing of data suggests that a much higher percentage of premature deaths are associated with obesity (over 18 percent) than previously thought.[14]

Studies have linked eating red and processed meat with an increased risk of developing colon and prostate cancer, poor metabolic health, and other diseases.[15] (But researchers have not conclusively proven that eating red meat *causes* cancer.) A Harvard School of Public Health study found that people who eat more red meat have a significantly greater chance of dying from all causes. As reported in the *Archives of Internal Medicine* in April of 2012, this study involved 37,698 men and 83,644 women and followed participants from 1980 to 2008 for a total of three million person-years. The researchers came to an attention-grabbing conclusion: *Eating one serving of unprocessed red meat a day increased a participant's chance of dying over the course of the study by 13 percent.* The chance of dying from cardiovascular disease increased by 18 percent, and of dying from cancer by 10 percent. If the red meat was processed (as in bacon and hot dogs), the chance of dying was 20 percent greater. The researchers took into account factors like a history of heart disease or diabetes, age, medication or supplement use, smoking, being fat, and being a couch potato (red-meat eaters are more likely to smoke, be fat, and be physically inactive).[16] Note that the study does not mean that the participants had a 13 percent chance of dying if they ate a serving a day of red meat; it meant they had a 13 percent *greater* chance of dying. If you eat red meat only every other day, the study suggests you cut that risk in half. Another finding of the Harvard study is that substituting healthier types of protein-rich foods for beef lengthens life spans. If participants had eaten only half as much red meat, 9.3 percent of the deaths in men and 7.6 percent in women could have been avoided.[17]

What's in beef that could cause so much harm? There appear to be many parts to the answer. Prime suspects are the saturated fat and the heme iron in red meat. Heme iron, found in animal proteins, is believed to be inflammatory.[18] The heterocyclic amines and polycyclic aromatic hydrocarbons that form on the surface of red meat cooked at high temperatures may cause colon cancer in people with certain genes.[19] And (in mice at least) diets like the Atkins diet that are high in protein from red meat, and low in carbs, can promote hardening of the arteries.[20] Processed beef (such as that in hot dogs and cold cuts) contains nitrates and nitrites (and sometimes nitrosamines) that can lead to cancer. Nitrates can also cause hardening of arteries. There's more salt in processed meats, too, and salt raises some people's blood pressure.[21]

A new theory is that a substance abundant in red meat, carnitine, is eaten by particular gut bacteria and turned into something nasty, a chemical called TMAO (trimethylamine-N-oxide). TMAO makes it harder for the body to excrete cholesterol and easier for cholesterol to bind to artery walls. The higher your blood levels of TMAO, the more likely you are to have a heart attack. In fact, TMAO levels predict your risk of heart attack better than cholesterol levels, although they are rarely tested.[22] This TMAO theory came as a surprise because carnitine is known to contribute to heart health in several ways.[23] It is a common dietary supplement—but now appears to be dietary candy for the bacteria that produce TMAO.

Mainstream medical opinion holds that substituting polyunsaturated fatty acids for saturated fat reduces your chances of having cardiovascular disease and is good for your overall health. Overconsumption of saturated fat is also a cause of the obesity epidemic. In response to a widespread campaign that began in the 1970s, Americans cut back on saturated fat, although most of the decline came from cutting back on eggs, not beef.[24] Beef and dairy are American's main source of saturated fat. Ground beef generally has more total fat than steaks, because it's not trimmed before being ground and extra fat might even be added.

Why is there so much saturated fat in our processed foods? One reason is that unsaturated fat spoils faster than saturated fat, so food processors tend to favor the latter; a longer shelf life means less waste. Another reason is that it tastes good. Manufacturers responded to the anti-saturated-fat campaign by removing or reducing the fat in their products. That made many foods taste like cardboard. So manufacturers added a lot of high-fructose corn syrup. Some controversial research hints that the increased amount of sugar in our diets (both fructose and sucrose) might not just make us fatter but might lead to insulin resistance and cancer.[25]

Grass-finished (or *USDA labeled* "grass-fed") beef comes from cows that since weaning have been fed only pasture, hay, and other grasses. All beef cows are grass-fed to some extent because they all spend time grazing, so without the USDA grass-fed logo "grass-fed" is meaningless. The vast majority of cows are finished on grain. Beef that's grass raised *and* finished can be so low in fat that it has only a wee bit more fat than a skinless chicken breast. Several studies have found that grass-finished beef also has a healthier balance of fatty acids than factory-farmed beef.[26]

Not worried about your heart or cancer? A recent study suggests that the saturated fats in red meat, cheese, and butter might also dim your brain. Harvard researchers collected information on 6,183 women over age sixty-five. Over a period of four years they periodically tested the women's memory and abstract thinking skills and asked detailed questions about their diets. The overall cognitive skills of those who ate the most "good" monounsaturated fat (from olive oil, canola oil, and safflower oil) actually improved over the course of the study. Women who ate the most saturated fat (such as beef, cream, and cheese), on the other hand, had declining verbal and global cognition. The study controlled for things like smoking, age, education, exercise, and medical problems.[27]

Unnatural Ingredients in Beef

Most Americans have some doubts about the safety of our food, and 16 percent are "not at all confident" about it.[28] We face escalating and poorly understood problems like autism, male birth defects, falling fertility rates, early onset of secondary female sex characteristics, children's brain cancer, and so forth, and we wonder whether there might be links to chemicals we're ingesting.

Much of our uneasiness stems from things that are either unintentionally or purposefully added to our food. We are the lab rats in a massive ongoing experiment—in spite of the billions we spend on federal organizations with names like the Environmental Protection Agency, the Food and Drug Administration, the Centers for Disease Control and Prevention, the Center for Food Safety and

Applied Nutrition, the Consumer Product Safety Commission, the Food Safety and Inspection Service, the National Institute of Food and Agriculture, and the National Institutes of Health.

Several times in this book we have criticized federal agencies that have failed to protect the public. That criticism is rooted not in anti-government cynicism but in clear-eyed idealism. We have personally known every EPA administrator since the agency was created, and we count several as close friends. Similarly, we have often watched regional EPA administrators and program heads act as profiles in courage, and we assume there are also many such people in other agencies. Our critiques are not a broadside against the government in general, and certainly not against the exemplary public servants who protect America from assaults on public health and the environment. They are heroes, often working against great odds, and we admire them enormously.

Still, something has changed in America in the last half-century, and not for the better. Bill Ruckelshaus, the courageous Republican attorney who served as the first head of the EPA under Nixon, and later brought back the agency from near-oblivion under Reagan, worries about this. Bill told us, "Back in the 1970s, other countries watched the way our science and public policy reinforced one another, and their environmental leaders rushed to our shores to marvel at our willingness to tackle tough problems. They studied our successes, learned from our mistakes, and followed many of our policies.

"We've lost some of that attractiveness because our domestic policies are not as enlightened, or innovative, or aggressive as they once were. We have broken down into squabbling camps. We've lost the willingness to see environmental problems as challenges that should excite or energize American ingenuity and optimism."

What happened? Provoked by some outrage, Congress will pass a strong law aimed at protecting public health. Then the policy pendulum swings back and a subsequent Congress will refuse to appropriate any money to implement the new regulations. The embattled agency then finds itself operating under a law that requires it to take action but with a budget that explicitly forbids it to spend money to implement the law. So when we mention the failure of an agency, the real fault often lies with some appropriation subcommittee in Congress.

Occasionally Americans elect presidents who are fiercely anti-government. The two years that Anne Gorsuch ran the EPA for President Reagan can be fairly characterized as a bureaucratic reign of terror. Gorsuch, with the president's blessing, did her best to dismantle the environmental agency. She drove out many of the best and brightest who had been recruited by such previous leaders as Bill Ruckelshaus, Russel Train, and Doug Costle, and she replaced them with zealots seeking to sabotage the agency's mission. For example, her assistant administrator responsible for hazardous waste was convicted of lying to Congress about the Superfund program and was sent to prison. Gorsuch herself was the first federal agency head in history to be cited for contempt of Congress. The anti-regulatory extremism in the Reagan and George W. Bush years often created intolerable working conditions for regulators who sought to follow the law.

Also, for compelling historical reasons, civil service employees enjoy job protection. In a dream world where this is combined with advancement based on performance, the policy might lead to Confucian perfection. In the actual world, it often allows risk-averse, low-energy clock punchers to rise through the ranks. The least able or the unwanted are sometimes relegated to bureaucratic backwaters where they vegetate. More commonly, they function like a huge wet blanket that thwarts young idealists. But when federal agencies are led by smart, courageous servants like Bill Ruckelshaus, Lisa Jackson, Don Kennedy, Luther Terry, and Tony Fauci—people who are able to attract to public service a corps of people in their own image and inspire jaded bureaucrats to recapture their youthful idealism—America works better than the nation's founders could have dreamed.

Since industrialization began, humans have deliberately created and released into their surroundings a hundred thousand previously unknown chemicals.[29] Additional chemicals are made accidently, as by-products of manufacturing or other processes. Cows' bodies and human bodies are full of chemicals that have not been adequately tested for their unintended impact. Singly, and in an infinite number of possible combinations and potencies, some of these chemicals might do great harm, even in tiny amounts.[30] They are found in many of our foods, not just beef and dairy. Residues of heavy metals, drugs, herbicides, and pesticides remain in our food after cooking; some may even be made more harmful by heat.

A key law passed to protect us from poisons is the Toxic Substances Control Act (TSCA) of 1976, which the EPA administers. Under this law, it's generally up to the EPA (not the manufacturer) to prove a chemical is unsafe. About 85,000 chemicals have now been registered, but that doesn't mean that they have been studied to see if they are safe, because the EPA grandfathered 62,000 preexisting chemicals. The EPA has not done even minimal independent testing on 95 percent of the chemicals in use, although many of them are very toxic.[31] Over nearly forty years, the EPA has issued regulations controlling the use of only five chemicals.[32] Therefore, activists dismiss TSCA as the Toxic Substances Conversation Act.[33]

The EPA, the USDA's Food Safety and Inspection Service (FSIS), and the FDA set tolerance levels for residual veterinary drugs, pesticides, and heavy metals. But the agencies don't set tolerance levels for many other toxic substances. If no tolerance level has been set, a toxin isn't tested. The EPA sets tolerance levels for various pesticides based on their risks to human health, and the FSIS enforces these tolerance levels for meat, poultry, and eggs. The FDA is charged with setting tolerance levels for poisonous or deleterious substances in all other foods, including dairy, for both animals and people. A fourth agency, the CDC, recently began testing people to try to find out how much of which toxins Americans actually have in their blood.[34]

Because the beef industry has such great influence on the federal agencies that are supposed to regulate it, environmentalists and consumer groups are deeply skeptical of federal assurances that growth hormones and pesticide residues in beef are harmless. An intriguing study published in 2007 by Shanna Swan at the University of Rochester Medical Center in New York fortifies such concern. The remarkably persuasive Ms. Swan coaxed 387 men born between 1949 and 1983 into giving her a sperm sample, along with information on how long it had taken them to conceive a child. She also asked the men's mothers how much beef they ate while pregnant with their sons. (A weakness of this study is that it's hard to accurately recall what one ate in 1949.) Swan found that men whose moms said they ate more than one serving of beef a day were more likely to have sons with a subfertile sperm count. About 18 percent of such sons had this problem, contrasted with only 6 percent of men whose moms ate only one or no portions of beef daily. Swan speculates that growth hormones given cattle, or the pesticides the cattle ate, might have interfered with the development of the sons' testes.[35] In an earlier study Swan wondered why semen quality was better in men from New York City, Minneapolis, and Los Angeles relative to men in Columbia, Missouri, a region where agricultural pesticides are heavily used. She looked at pesticide metabolites in the urine of fifty Columbia men (twenty-five with normal sperm and twenty-five with unhealthy sperm) and confirmed that greater exposure to nonpersistent pesticides was associated with less healthy sperm.[36]

Many federal agencies have a semi-independent Office of Inspector General (OIG), whose duty it is to monitor that agency's work for effectiveness, waste, fraud, accounting shenanigans, and much else. The annual reports issued by OIGs offer a wealth of information that is otherwise difficult for the public to obtain. The USDA's OIG issued a report in 2010 on the FSIS, which samples meat for residual veterinary drugs, pesticides, and heavy metals. The OIG looked at how well the agency was doing this and concluded it was failing. According to the OIG report, the FSIS relied on "often antiquated and ineffective" testing methods. The FSIS approach of having workers put pen-and-paper notes on tags attached to carcasses almost seemed designed to fail. If a cow carcass tested above the limit for a toxin, the FSIS couldn't prosecute producers who violated standards, because it generally had no way of knowing where the cow came from.[37] (The USDA can't force a recall of dangerous meat. All meat recalls are voluntary, although companies have always thus far complied with requests.)

A new computer system was installed to help meat inspectors file their reports, and to get samples to labs and the results of lab tests back more quickly. The computer now tells inspectors which meat to sample and when to sample it. A March 2013 OIG report says it's working poorly, however. The computer keeps crashing, and a hundred million pounds of beef were sold to consumers without being properly sampled.[38]

About 80 percent of cows in American feedlots are given supplemental hormones to make them grow faster.[39] This increases profits and lowers the cost of beef. The practice began in the mid-1950s. Three of the hormones used in beef cows occur naturally in both humans and cows; three are synthetic but mimic the effects of naturally occurring hormones. Some of these hormones are the same ones cheating athletes use to enhance their muscle mass and performance. (Not everyone agrees that the hormones boost producer profits. Journalist and author M. R. Montgomery, who has much firsthand experience with cattle, argues: "The few studies not done by hormone-implant manufacturers suggest the cost-benefit ratio is a wash."[40] Oklahoma cattleman David Sheegog claims that hormone-stimulated beef is 25 percent tougher, and that the higher value of tender beef more than makes up economically for the slower weight gain.)[41]

The FDA says meat from steroid-hormone-treated cows is safe for humans to eat at any time after the animal is treated.[42] The Scientific Committee on Veterinary Measures Relating to Public Health of the European Commission disagrees. They ran experiments and found that the injection of hormones increases the level of estrogen residues in meat seven- to twentyfold. The committee also maintains there are poorly understood risks of "neurobiological, developmental, reproductive and immunological effects, as well as immunotoxicity, genotoxicity and carcinogenicity," and that it's impossible to establish safe threshold levels. Therefore, the European Union bans most uses of hormones in beef cattle. Furthermore, the European Union doesn't allow meat to be imported if it contains artificial beef hormones.[43]

Some believe the European ban was driven more by public alarm than by good science. Here's the background: In 1977, schoolchildren in northern Italy showed signs of a premature onset of puberty and eighty-three schoolboys grew breasts. Beef or poultry in school lunches—meat from animals given growth hormones—was suspected, but not confirmed.[44] Unfortunately, no allegedly tainted meat was still available to be tested.[45] In 1980, there was another scare when an Italian consumer group found illegal DES in French veal baby food. It was claimed that boys and girls who ate the baby food grew breasts, and girls to began to menstruate.[46]

The European Union banned imports of American beef produced with growth hormones, and the great trans-Atlantic food fight began. The United States viewed the ban as protectionist and demanded sanctions. The Reagan administration slapped tariffs on $100 million of European exports. When new rules on international trade became available, the United States challenged the ban under the World Trade Organization (WTO). The WTO Appellate Body (a small group of lawyers and academic economists but no scientists) said the ban on the importation of beef treated with growth hormones was irrational. Europe immediately appealed the WTO ruling. In March of 2010, the European Union agreed to allow imports of a larger amount of hormone-free American beef, and the Americans agreed to lift tariffs on certain European foods.[47] But the hormone ban endures.

When anabolic steroid growth hormones were banned in Europe, some farmers on both sides of the pond turned to beta-agonist drugs to make cattle quickly put on lean weight prior to slaughter. The FDA says it's safe to use these drugs.[48] American beef producers, hit with high feed prices because of drought, love beta-agonists because they can add thirty pounds of beef to a steer, another thirty pounds of profit.[49] The drugs quickly break down, and are said not to accumulate in cows' meat.[50]

Some health experts say the impacts of these drugs on beef-eating humans aren't clear, and that beta-agonists may cause lameness, skittishness, and heart problems in cows.[51] If residues are left in beef they may cause problems in humans: tachycardia, tremors, nervousness, muscle aches, and headache.[52] Merck responds that over thirty studies have found no safety issues with their drug.[53]

Beta-agonists have been banned from use in cattle in Europe, China, and Russia, but are legal in Canada, Australia, and the States. Until mid-August 2013, it was estimated that over 70 percent of American beef cattle were given the beta-agonists Zilmax (nicknamed "Vitamin Z") or Optaflexx prior to slaughter. Then a video of cows limping on their way to be killed made the news. Tyson Foods (which slaughters one out of four American cows) quickly announced it would stop accepting cows fed Zilmax. Merck, the maker of the drug, suspended production. In December 2013 Merck announced it would resume supplying the drug to "certified" feedlots. To get certification, Merck said they would require a feedlot's veterinarian, nutritionist, and management to go through training about Zilmax's proper use. Studies will be done of Zilmax's effect on cow motility.[54]

Optaflexx sales boomed after Zilmax sales were suspended. Here's the warning on Optaflexx:

> WARNING: The active ingredient in Optaflexx, ractopamine hydrochloride, is a beta-adrenergic agonist. Individuals with cardiovascular disease should exercise special caution to avoid exposure. Not for use in humans. Keep out of the reach of children. The Optaflexx 45 formulation (Type A Medicated Article) poses a low dust potential under usual conditions of handling and mixing. When mixing and handling Optaflexx, use protective clothing, impervious gloves, protective eye wear, and a NIOSH-approved dust mask. Operators should wash thoroughly with soap and water after handling. If accidental eye contact occurs, immediately rinse eyes thoroughly with water. If irritation persists, seek medical attention. The material safety data sheet contains more detailed occupational safety information. To report adverse effects, access medical information, or obtain additional product information, call 1-800-428-4441.[55]

Bison vs. Cows

If you crave red meat, consider bison. It's illegal to give growth hormones to bison and they get antibiotics only if they are sick, which they seldom are. Bison have deep North American roots. They belong here. Bison has a sweet, rich flavor, and steaks are tender if you don't overcook them. A hundred grams of what the USDA calls "separable lean only" raw bison has 109 calories and 1.8 grams of fat

The original range of Great Plains bison (darkest color) is far greater than most Americans realize. From Wikipedia, http://en.wikipedia.org/wiki/File:Bison_original_range_map.svg. Source: Simon Pierre Barrette, CC-BY-SA, via Wikipedia Commons.

in contrast to "separable lean only" Choice-grade beef's 291 calories and 24 grams of fat.[56] Look for grass-finished bison.

The prairies weren't empty when the cows began moving in. The long- and short-grass prairies were home to vast herds of bison that had co-evolved for a million years with the plants that fed them.[57] At one time there were an estimated twenty to sixty million of the shaggy beasts with horns like those on Viking helmets. They roamed eastern forests and the Gulf Coast as well as the prairies, covering just about all of the future United States except the East and West coasts. Today, there are about half a million bison in North America, most of them in private herds.

Bull bison are the largest existing animals native to North America. Males weigh up to a ton. Bison are strong and aggressive, can outmaneuver horses, break through ordinary barbed wire fences as if they were cobwebs, jump six feet up into the air, and run thirty-five miles per hour. It's hard to say how smart bison are; if you shoot one, other bison standing nearby will go right on grazing. But their sense of smell and their eyesight are excellent. Adult bison are not gentle. Their big black tongues are . . . not cute. When shedding hair in the spring and summer adult bison look ridiculous. (The golden babies, however, are adorable.)

The real problem with bison, from the European settlers' point of view, wasn't their disposition or appearance, but that they fed and clothed and housed the problematic native peoples. Settlers made heroes of men who shot and killed bison, men like William Cody, aka Buffalo Bill. Cody killed 4,280 bison over eighteen months to feed men building a railroad. A hero to many, he then toured with his "Wild West Show." Cody was far from alone. The great bison slaughter occurred over only three years, from 1872 to 1874. Over the three years of the great slaughter 1,378,359 bison hides were reportedly tanned.

When it was over, only a few hundred bison remained. Aside from a couple dozen in Yellowstone National Park, the survivors were the semi-wild pets of a handful of men, white and native, who cared about them. This unprecedented slaughter was no accident. It was government policy. General Philip Sheridan, who then led the Military Division of the Missouri, praised the great buffalo wipeout, and the true reasons for it, with astonishing candor: "For the sake of a lasting peace [with the Indians], let them kill skin and sell until the buffaloes are exterminated. Then your prairies can be covered with speckled cattle and the festive cowboy."[58]

Bison were also slaughtered because there was a strong market for their hides. Europeans had been enjoying a flood of cowhides from Argentina and had gotten used to having nice leather products. Europeans discovered that bison hides were even more supple than cattle hides, but the cattle introduced to the South American pampas had destroyed that environment such that it would no longer support vast numbers of cattle. (This is eventually the fate of all open-range endeavors where grass is free and its use unregulated.)

Concern about endangered species was a century away. Even today, the idea of letting Indian tribes keep bison gives some cattle ranchers the willies. They don't want free-roaming bison competing for grass with their cows and providing reservoirs for bovine diseases. They aren't even keen to allow tribes to keep buffalo on their own reservations.[59] *New York Times* reporter Jim Robbins quotes a Montana state senator as saying, "Why do you want to spread this creeping cancer, these wooly tanks, around the state of Montana?" Further mixing his metaphors, the senator compares buffalo to dinosaurs. He apparently doesn't want dinosaurs in Montana, either.[60]

• • •

After the great slaughter of 1872–1874, bison skulls were piled up, waiting to be ground for fertilizer circa 1875. Photo: courtesy of the Burton Historical Collection, Detroit Public Library.

Like cows, bison are in the bovine subfamily. But domestic cows are *Bos taurus*, and the bison in North America are *Bison bison* (there is one other bison species, in Europe). But wait a minute: Isn't the definition of "species" a population of critters capable of interbreeding? And aren't there creatures called "beefalo," which are three-eighths bison and five-eights cow?[61] And "cattalo," which are hybrids that look like bison? Must have been some interbreeding there.

Well, it's complicated. The Bovinae subfamily split-up happened a million years ago. No one knows why. Left to their own preferences, modern cattle and bison don't mix it up romantically, but with human assistance offspring can be produced. At first, most offspring proved to be sterile. But after experiments with cattalo that couldn't reproduce well, ways were eventually found to make fertile beefalo. Since then, a big family reunion has been under way, leaving few herds that are pure bison (the herds in Yellowstone and Wind Cave national parks might be free of cattle genes). Nearly all state and private herds are "contaminated," but most of the "contaminated" herds don't carry many cow genes, so they should still be able to adapt pretty well to prairie living and fight off wolves better than cows can.[62]

As today's biology students know, there are at least two types of inherited genes (this is as true for you as it is for cattle and bison). First, there is the familiar nuclear DNA that comes from Mom and Dad and forms the human genome. Then there is the DNA hidden away inside the little powerhouses (the mitochondria) inside every cell of your body, and that DNA comes only from Mom. It's called mitochondrial DNA (mtDNA).

Bison mtDNA has recently been sequenced, so comparisons among various herds and between cows and bison are possible. Scientists have already taken a peek at the mtDNA of federal bison herds and have found almost no domesticated cow mtDNA in those animals. The reverse is true in state-managed and private herds.

So what does it matter? Don't beefalo make fine burgers? Like bison, beefalo can also be lower in fat, calories, and cholesterol than beef.[63] (Look for grass-raised-and-finished beefalo.) But true bison have a very, very long history of surviving population collapses and being able to handle extreme changes in climate, population fragmentation, imported *bos* diseases, "bison-cide," and destruction of their habitat.[64] In short, they have amazingly good genes—a pedigree that means something and that we tamper with at our risk. It is very much in our self-interest to maintain these blueblood bison. There's nothing *wrong* with beefalo. Just keep them away from the bluebloods.

Because bison and the plants and wildlife of the prairies co-evolved, bison can find their own food. They are adapted to life on the Great Plains in a way cows never will be. It isn't necessary to grow hay, corn, or other grains to feed and fatten bison (although some buffalo ranchers do this). Dan O'Brien, a former cattle rancher who switched to bison, has written a convincing book, *Buffalo for the Broken Heart*, in which he describes how his bison herd helped restore the grasses, ponds, and wildlife on his South Dakota ranch. Unlike O'Brien's former herd of cows, bison don't need help giving birth, require far fewer medications, and can survive on less water from more obscure sources. They can endure cold winters outdoors because they are better insulated: Compared to cows, they have ten times as many primary hairs per square inch, and their shorter eyelashes discourage ice buildup. If you ever get a chance, bury your face in the thick winter pelt of a bison. At least run your fingers through it. It's soft as snow, yet warm as the sun on your back.

Bison can plow their own way through snow. They prefer drier grass than cows, and rougher roughage. They can run farther and move frequently as a herd to new pasture before the old one is picked bare.[65] O'Brien says his bison don't spread out like cattle when they graze but naturally stick together in a bunch (a trait probably developed as protection from wolves), and graze intensely, their hooves driving seeds into the soil.

But raising bison requires sterner stuff than can be bought at a western-apparel shop. Ranch hand and writer Bryce Andrews once worked buffalo at the Ted Turner ranch.[66] He told us he found the job "bloody and terrifying." Once or twice a year the creatures need to be rounded up for vaccinations or shipping. They don't much like it and gore one another with their horns. Four bison died in just one day from goring. The Turner ranch, Bryce said, was staffed with experienced people and had high-quality equipment and sturdy metal fences. He told us the story of a guy who bought four bison bulls and loaded them into an aluminum trailer; horns soon began popping right through the sides. "They opened it up like an aluminum can."

(When bison are raised in crowded conditions, some ranchers poll them or cut off the tips of their horns. Others object to polling bison; it decreases their ability to defend themselves in the wild. Many object to raising bison as if they were cows.)

Kansas State University is running a long-term, side-by-side experiment on a large tall-grass prairie reserve, the Konza Prairie Biological Station. On one chunk of prairie are domestic cattle, on another are three hundred bison. The researchers are studying how bison, cows, climate change, and fires affect the prairie ecosystem. Both cows and bison on the reserve eat mostly big bluestem, little bluestem, and Indian grass. The bison-inhabited prairie has developed greater plant diversity; more forbs grow in the bison prairie. The bison are pretty much left to fend for themselves year-round. During the winter, they lose about 11 percent of their summer weight, and they put on weight more slowly than domesticated cattle. The cattle are much higher-maintenance: If left alone and not fed supplementary food, they would die during the winter.

Sentiment might now be shifting in bison's favor. When the United States mint decided to issue a special quarter for every state (between 1999 and 2009), three states featured a bison on their quarter: Kansas, North Dakota (two bison), and Montana (a bison skull). Two pictured cows: Nebraska (oxen pulling a covered wagon) and Wisconsin. The National Bison Legacy Act would declare the bison to be the national mammal. Shortly after the bill's introduction in 2012, a white bison bull calf was born in, of all places, Connecticut.[67] Some Native Americans see this as an extremely auspicious sign. The bison is named Yellow Medicine Dancing Boy, and his bison bloodlines are pure.[68]

Robust sales of bison meat have defied the current economic downturn. There is more demand than supply, and that situation seems likely to continue. The National Bison Association (NBA) believes bison will remain a small part of the red-meat market. "Bison producers never want our product to be a commodity," the NBA says on its website.

We'll always need cows. Cows are essential to dairy; female bison don't take kindly to humans handling their small teats ("buffalo mozzarella" comes from water buffalo, not bison). But given the weather that might be headed toward the plains, and given the stresses on the Ogallala Aquifer, North Americans would be wise to raise more bison and fewer cows. We're lucky to still have a bovine available to fill an ecological niche that might be difficult to put to other uses.

Meat Snubbers

The bite of a tiny insect can force some people to stop eating beef. Lone star ticks (common ticks with a white star or dot on their back) spread a disease called galactose-alpha-1,3-galactose ("alpha-gal") that makes people allergic to mammal meat. Aside from alpha-gal, mammal meat allergies are rare. Victims of alpha-gal may get hives, uncontrollable itching, and feel sick to their stomach. But even worse, a lone star tick bite can cause delayed anaphylaxis—after eating mammal meat people may feel okay for a few hours but then go into shock.[69] Best-selling mystery author and lawyer John Grisham is one victim of alpha-gal. In 2002, Grisham's skin (even on his ears and ankles) episodically began to feel as if it were on fire. It formed welts and itched. Trying to solve his personal mystery, he kept track of everything he ate and eventually connected the itching to eating beef four or five hours before the problems began. He gave up eating beef.[70]

According to a Public Policy Polling survey, about 13 percent of

Americans choose not to eat any kind of meat, poultry, or fish. An additional 7 percent are vegan and consume no animal products, not even eggs or dairy.[71] (A word of caution: The results of polls on vegetarianism differ greatly.) A conundrum for lacto-ovo vegetarians, who consume dairy and eggs but not meat, is what society should do with a cow or chicken after it stops producing.

People become vegetarians for various reasons. Slightly more than half of vegetarians, including former president Bill Clinton, reportedly chose their diet for health reasons.[72] Young people are more likely than older people to embrace vegetarianism because of concern about cruelty to animals. Although it is not a stampede, in America there is clearly a trend toward eating less meat.

Lab-grown "real" meat has become a reality—on a teensy scale—and purports to address all the issues that persuade people to become vegetarian. In August 2013, in front of an audience at a London theater, two tasters chomped down on a lab-grown hamburger. The constrained politeness of their evaluations and the expressions on their faces said it all: Although the burger contained salt, breadcrumbs, and beet juice to disguise its grayish-yellow color, the total lack of fat diminished its appeal. Nor did it contain any iron.

Mark Post, a tissue engineer at Maastricht University, Netherlands, created the five-ounce cultured burger at a cost of $325,000. The small, very thin sheets of meat have already been nicknamed "shmeat." Proponents of the process prefer to call the process "in vitro" and the product "cultured beef." This burger was grown from stem cells taken from cow shoulder muscle and multiplied in a medium of fetal-calf serum obtained by killing pregnant cows. The plan is to eventually replace cow-derived materials with materials that don't come from animals, and to find a way to culture the right sort of fat cells. Post, whose work was funded by Google co-founder Sergey Brin, thinks the product might be mass-produced in a decade or two.[73]

Scientists at the University of Oxford and the University of Amsterdam argue that if we could grow only the parts of cows we want to eat, rather than the entire animal, it would dramatically reduce the environmental impact of cows. Although they admit to much uncertainty in their calculations, they contend that cultured meat would use 7 to 45 percent less energy, produce 78 to 96 percent fewer warming gases, consume 82 to 96 percent less water, and require 99 percent less land.[74] In addition to these possible environmental benefits, people who don't eat beef because they object to mistreating and then killing sentient creatures might enjoy lab-grown meat. The lab meat could be grown in sterile conditions and "to order."

Professor Gabor Forgacs of Clarkson University is also interested in cultured meat. His technique uses 3-D bioprinting. You can watch him eat a sliver of his lab meat during his TEDMED talk on the Web. He first fries it in some oil and adds salt and pepper.[75] Billionaire-iconoclast Peter Thiel, perhaps best known for funding exceptionally smart students who agree to drop out of college to launch companies, funds Forgacs's company, Modern Meadow. The firm intends to start with leather, not meat, because it might be easier for people to accept something they sit on than something they ingest. The company's CEO, Andras Forgacs (Gabor's son) pitches the whole concept in terms of diminished environmental impact in a world with limits. He closes by comparing the gruesome awfulness of even the best slaughterhouse to the cruelty-free, lablike ambience of a cultured meat plant. He suggests companies may open restaurants associated with their meat-growing operations "and operate like microbreweries," with the creation process on display behind a window.[76]

There are obstacles to this dream of lab-grown meat. If the energy to grow meat doesn't come from the sun (and hence into grass and then into cow), that energy has to come from somewhere else. Stem cells must be fed to grow. True, you could grow just steak and not horns, hooves, brains, and big brown eyes, so there would be savings there. Researchers Hanna Tuomisto and Avijit Roy argue that if all the meat (not just beef) produced in the European Union was cultured, greenhouse emissions and land and water use could be reduced by two orders of magnitude. By far the largest reductions were predicted in the production of beef, as opposed to other types of meat. They assumed stem cells would be taken from animal embryos and cyanobacteria hydrolysate would be used as the nutrient to grow muscle cells. Their study did not consider the energy needed to exercise the growing muscle cells, or the resources needed to replace meat by-products like leather and wool, and the researchers acknowledge many other uncertainties.[77]

Molecular biologist Mardi Mellon of the Union of Concerned Scientists is skeptical. She told NPR reporter Ketzel Levine that lab meat would be bad for the environment: "Picture it: You've got a big compound of buildings with scientists running around tending big vats of cultured cells, making sure that they're all at a constant temperature, that the cells are being kept sterile. . . . I mean, where does that energy come from?"[78]

Thermodynamics offers no free lunches. But if cultured meat is grown in solar-powered bio-reactors in a bath of nutrients, it might, in theory, supply protein while reducing the carbon footprint of beef and eliminate the cruelty now endemic in its production. Deep thinker and author Steven Pinker argues that fake meat will be the final step in the "rights revolution"—the decline in human cruelty that he argues has taken place over the past few centuries.[79]

Chapter Seven

CANNIBAL COWS

In December of 1997 we were in Stockholm because Gail's father, Paul D. Boyer, had won a Nobel Prize in Chemistry. That same year, Stanley Prusiner won the Nobel Prize in Physiology or Medicine for his work identifying the agent that causes mad cow disease. A tall, handsome man with excellent posture and a mane of curly, silver hair, Prusiner strode through the public rooms of the Grand Hotel, where Nobelists are lodged, with acolytes in his wake. In his Nobel lecture, Prusiner explained how a misshapen protein, something that is not "alive," is nonetheless contagious. He had dubbed this infective agent a "prion" (PREE-on, derived from the words protein and infectious). When people eat meat from infected cows, they swallow infectious prions. Those who are susceptible then develop an incurable, fatal disease that eats holes in their brains. A powerful emotion, possibly rooted in afternoons in darkened theaters watching 1950s science fiction movies, stirred when we heard this. We perked up and paid close attention.

Once initiated into the then tiny fraction of the public aware of prions, over the years we continued to read up on them and follow the news on outbreaks of mad cow disease. We think that the United States remains vulnerable to a costly epidemic that could cause considerable human misery.

Since the 1700s, a mysterious disease in sheep and goats that makes their brains look like sponges has been recognized. Called scrapie, it causes infected animals to obsessively scrape off their pelts, develop a funny gait, and die. Early in the twentieth century, German scientists described a disease in humans they dubbed Creutzfeldt-Jakob disease (KROITS-feld YAW-kob). CJD victims lose their minds along with control over their speech and muscles, and then die. Other (exceedingly rare) prion diseases also affect humans: kuru, fatal familial insomnia, and Gerstmann-Sträussler-Scheinker syndrome. In animals there is chronic wasting disease in North American deer and elk, and similar diseases in cats, mink, and some zoo animals. Before it was known that prions existed, all such diseases were categorized as transmissible spongiform encephalopathies (TSEs). In layman's language, TSE means "infectious disease that makes brains look like sponges." They are all fatal diseases.

Scientists now know that not all prions are infectious and that normal prions are found in the cell membranes of healthy mammals as well as in some other life forms. Prusiner was initially shocked to learn this. But he was able to show that prions in diseased creatures had a different three-dimensional conformation. Normal prions may play a role in long-term memory, in helping brain cells communicate, and in renewing bone marrow.[1]

For reasons still poorly understood, a prion sometimes becomes misfolded in a manner that allows it to replicate itself (to become infectious) and to protect itself from enzymes that normally break down proteins. This misfolded prion acts as a template that triggers a nearby normal prion to assume the misfolded shape—turning it from Dr. Jekyll into Mr. Hyde. Newly refolded prions then reshape still more prions. Clumps of these form amyloid folds (deposits of proteins outside of cells) and disrupt the structure of the brain. The mystery of exactly *how* the misfolding of prions is induced hasn't been solved.[2]

Here's another really scary thing about prions: they survive cooking, disinfectants, radiation, rendering, sewage treatment plants, composting, being buried for five years, and even the medical sterilizing devices called autoclaves! Remaining viable at temperatures as high as 600 degrees Celsius (1112 degrees Fahrenheit), they make infectious bacteria, fungi, and viruses look like wimps.

A few years after our Nobel winter, we made the road trip around the British Isles mentioned at the start of this book. Britain was still reeling from its horrific outbreak of mad cow disease. The year of our travels, it claimed the lives of eighteen more people. As of February 2014, there were 177 definite or probable human deaths in the United Kingdom, 27 cases in France, and a scattering of cases in ten other countries. Fortunately, the number of newly diagnosed cases has fallen off sharply, although some scientists expect another burst or two for reasons explained later.

We wondered why a disease as dreadful as mad cow went undiscovered for so long in a scientifically advanced nation. Why had it already infected hundreds of humans before it was officially acknowledged as a threat? One reason, we learned, is that the incubation period in humans can be surprisingly long, making it difficult to link the illness to a hamburger eaten years, or even decades, earlier. (The final death count in Britain might not be known for fifty years.)[3] Another reason, however, was officialdom's desire to protect a valuable British industry. The few who knew or suspected the truth hid it from the public. The same impulses that caused British officials to lie might motivate American officials to do likewise.

The epidemic began stealthily, just before Christmas in 1984. At Pitsham Farm in West Sussex, southern England, the behavior of cow number 133, normally a sweet cow, became peculiar. She was arching her back, waving her head back and forth, shaking, and falling.[4] In April 1985, in the nearby county of Kent, a Holstein dairy cow named Jonquil, who had always been gentle and friendly, began shying away from people. She stumbled around, bashing other cows, and soon "went down" and was killed and buried.[5] When cow number 133 died in 1985, her brain was autopsied and found to be full of holes. But the discovery went largely unremarked at the time.

Cows in other herds in Cornwall, Devon, and Somerset counties began to show similar symptoms. They kicked at people trying to milk them, held their heads too low, and lost coordination and weight. They shied away from crossing concrete and from going around corners or through doorways. They raised their hind legs too high when walking. One cow even chased a man on its knees.[6]

Veterinarians were baffled. The cows were of different breeds and widely dispersed, so it seemed unlikely the disease was directly transmitted from one cow to another. Beef cattle seemed unaffected.[7] (Of course, dairy cows are eventually slaughtered and their meat sold for human consumption.)

Researchers autopsied more cow brains; they all looked similar to the brains of sheep and goats that died from scrapie, a disease not believed communicable to either cows or people under ordinary conditions. But even as early as the autopsy of cow 133's brain, a few scientists thought they were looking at a new disease, which they called bovine spongiform encephalopathy (BSE). The normal procedure would have been to publish reports in scientific journals, which would have led to a faster understanding of the disease and more information on how it was spreading. But top officials in the veterinary investigation service and the Ministry of Agriculture decided not to release any information because they believed that doing so could "lead to hysterical demands for immediate, draconian government measures and could lead to a rejection of UK exports."[8] This was just an early example of many government efforts to conceal what was happening.

New vaccines, chemicals, and changes in cow genetics were investigated and ruled out as sources of BSE. Researchers then looked at food fed to dairy cows, but not to beef cows. Making huge volumes of milk is hard work, so today's high-producing dairy cows usually need protein supplements. Much of that protein came from animals: unwanted parts of dead cows, sheep, pigs, and chickens all went into renderers' stews. Fat was extracted from the stew, and the rest dried and made into stable products like bone and protein meal. Cows had been eating rendered food for decades with no problems. Had anything changed in the rendering process?

It had indeed. Many renderers had adopted a new, lower-temperature process, and many had also stopped using a solvent that extracted fat; fat acts as a protective shield for microbes. Although it was not certain that these processing changes were responsible for the outbreak (and it was later learned that microbes weren't responsible for mad cow disease and that even the old rendering method wouldn't have destroyed prions effectively), a veterinary epidemiologist led a study that confirmed that protein supplement was the source of BSE.[9] Cows had recently begun eating even more rendered cow parts when the prices of two other protein-rich foods, soy and fishmeal, rose.[10]

There is still disagreement over where the abnormal prion(s) originated. One theory is that the cows caught BSE from rendered sheep infected with scrapie. But many other scientists believe the disease arose spontaneously in one or more cows, cows that were rendered and fed to other cows. Those other cows became sick and were in turn rendered, and the disease spread rapidly.[11]

In July 1988, the British government banned the feeding of ruminants (hoofed herbivores that chew cud) to other ruminants. The government did not, however, ban feeding beef from infected herds to humans.

In May 1990, in Bristol, a Siamese cat named Max began behaving oddly and staggering. He was euthanized. In the following few years, more than one hundred cats would show similar symptoms and die of what came to be called feline spongiform encephalopathy.[12] Officials reluctantly admitted the cats had probably eaten infected beef in cat food. A wide variety of zoo animals also died from TSEs after eating rendered supplement. This cross-species spread of a TSE was something new, and hard to accept.

To reassure Brits that eating beef was safe for humans, the Conservative head of the Ministry of Agriculture, Fisheries and Food (MAFF) stood in front of television cameras on May 16, 1990, and in an exchange that went viral, offered his four-year-old daughter a burger. She thrust it away. The minister himself chomped down and swallowed it.[13] He claimed he had the clear support of the Department of Health and of "scientists who deal with these matters." This implied a unanimity among scientists that didn't exist. In fact, in 1988 the previous head of MAFF had been warned by his animal health division that there was some risk to human health and that the ministry should face the fact that the disease could spread to humans. For eight years this opinion was kept secret.[14] Fifteen years after the photo op, the daughter of a close friend of the minister was diagnosed with the disease on her twenty-first birthday. According to her father, "she was more helpless for those last two years than when she was born—at least then she could move her arms and cry, but by the end she couldn't even do that."[15]

In 1993, two dairy farmers died after showing TSE symptoms. The government told people not to worry, that the farmers had died of sporadic Creutzfeld-Jakob disease (CJD), long known to affect a very few elderly humans. But that same year lively, freckled, blue-eyed Victoria Rimmer, a resident of Wales who was only fifteen and who loved horseback riding, sports, ballet, and beef burgers, began feeling tired all the time. She began forgetting things. Formerly a teen who always had a smile on her face, she became moody and depressed, lost weight, and began staggering around and falling.[16] Nevertheless, eight doctors said Vicky was just fine. She lost her sight. A brain biopsy (a procedure with serious risks done on live people) revealed Vicky had a TSE. Nine cases similar to Vicky's turned up in other people while she was in the hospital. Many victims were young. (It's unknown why the disease favors young people. Some speculate it is because young people eat more beef burgers and are more likely to have tonsillitis or gastroenteritis, which may cause broken membranes, allowing prions an easy way into their bloodstreams.)

In March 1996 British officials finally announced that a new form of CJD had been discovered that was probably caused by eating beef. Because slices of BSE brain from infected cows look remarkably like slices of brain from humans who die of CJD, the new disease in humans was named *variant* Creutzfeldt-Jacob Disease (vCJD). But as today's CDC website acknowledges, "There is strong epidemiologic and laboratory evidence for a causal association between variant CJD and BSE."[17] There has never been a confirmed case of vCJD in any region free of BSE. So BSE in humans should really have been named vBSE.

As the government had feared, panic spread. Other nations banned the import of British beef. All cattle over thirty months old were ordered killed. Their remains were fed into incinerators that worked around the clock. In the countryside, power shovels lifted dead cows above great piles of their kin, and dropped them. As the bodies burned and billows of sooty clouds rose, delicate-looking cow legs stuck out of the piles at random angles like pins out of giant pincushions. Farmers and their families wept. Many lost their livelihoods forever.[18] Although 179,000 cows were known to have had mad cow, it was necessary to slaughter and burn or bury *4.4 million* cows to stop the epidemic and restore public confidence in beef and dairy foods.[19] (Milk and other dairy foods were apparently always safe to consume. No scientific evidence has come to light that shows cow milk to contain infectious prions.)

Vicky Rimmer finally died in 1997 after spending her final years in a coma. Her inquest was delayed for four years, until 2001. Even though everyone knew by then that people could catch mad cow disease, the coroner concluded that Vicky had likely died of bronchial pneumonia caused by sporadic CJD, and not from vCJD. The urge to suppress bad news had outlived the need.

Most people who ate infected beef would not catch the disease, but the extremely long incubation period and varied symptoms fueled anxiety. An animal or human's genetic makeup helps determine its vulnerability to particular infectious prions. Some breeds of cows appear more susceptible to BSE than others, as are some humans.[20] The human prion gene comes in two forms: M and V. That means three combinations are possible: MM, VV, and MV. The incubation periods vary for each combination. So far, only people with the MM combination have come down with mad cow. This was also true of the early cases of kuru (a disease found in a tribe in New Guinea probably caused by people eating the brains of their infected, deceased relatives). But a rash of later kuru cases were all in people with the MV combination.[21]

There is no practicable way for people to find out if they are infected with mad cow. Depression is a common early symptom in humans. Some people have hallucinations, develop odd phobias, or have strange sensory experiences, such as "sticky" skin. Victims get wobbly on their feet, become demented, their muscles jerk, and they have seizures. Late in the disease people stop moving and lose their ability to speak.

Eating infected beef isn't the only way of getting a TSE disease.

Sometimes a "sporadic" or "spontaneous" change seems to turn a normal prion into an infectious prion and sets off a chain reaction.[22] (Of course, nothing happens without a cause. It's just that the cause of sporadic TSEs is unknown.) For example, most cases of CJD are sporadic. A very few people develop CJD solely because of an inherited mutation in their prion gene, and USDA-affiliated researchers have discovered that at least one form of BSE is hereditary in cattle.[23] The take-home lesson: Although the spread of mad cow disease can be *contained* by not feeding diseased cows to healthy cows, hereditary and sporadic forms of mad cow disease will still occur.

Concealing the outbreak for so long magnified the harm it caused in Britain. Nearly thirty years passed from the time the first spongiform cow brain was found until the United States lifted its ban on the import of British beef in November 2013. But in December 2013, Australia and New Zealand—nations with valuable beef industries that have had bans on importing British beef in place since 1996—expanded their bans to include beef from all EU countries.

Impact of the UK Mad Cow Outbreak on the United States

Shortly after the mad cow epidemic was acknowledged by British officials, but before a single mad cow had been found in the United States, Oprah Winfrey interviewed Howard Lyman, a fourth-generation Montana rancher, on her television show. As reported by Sam Howe Verhovek in the *New York Times,* Lyman told Winfrey that cows in America were still fed ground-up animal parts, including parts of dead cows.

"But cows are herbivores!" Winfrey exclaimed. "They shouldn't be eating other cows."

"We should have them eating grass, not other cows," Lyman said. "We've not only turned them into carnivores; we've turned them into cannibals."

Then Winfrey said: "It has just stopped me cold from eating another burger."[24]

Beef prices fell for two weeks. Texas cattlemen sued Oprah for $12 million plus punitive damages under the state's new False Disparagement of Perishable Food Products Act (1995). Before the case went to the jury, the judge ruled that cattle were not a perishable food, as defined by the statute, and that Oprah had not disparaged any of the plaintiffs by name, nor Texas. The jury found for Oprah. The cattlemen appealed, but the Fifth Circuit upheld the lower court.

Thirteen states now have food libel laws similar to the Texas law. Under normal libel laws, the person who thinks he or she was libeled must prove that the offensive statement was false. Food libel laws often shift the burden to the defendant to prove that what he or she said about a food product was both true and based on reliable science. Moreover, some food libel laws award attorney fees and punitive damages to those bringing the lawsuits, if they win, but not to defendants who prevail. These industry-sponsored laws create an uneven playing field—the well funded can sue shallow-pocketed public interest groups. The constitutionality of these laws is questionable, but their effect on chilling discussion and debate is clear.

Four mad cows have now been officially acknowledged in the United States. Four persons with vCJD have been found. According to the CDC, there is strong evidence that three of those infections were acquired abroad. The fourth case, a patient who died in Texas, had traveled to Europe and the Middle East, but as of September 2014 it remained unclear where that person contracted the disease.[25] In Canada, as of April 2012, nineteen mad cows had been found. Canada has also had two confirmed cases of vCJD (one in 2002 and one in 2011). Cows, beef, and cow products move quite freely across the United States/Canadian border.

The first mad cow found in the United States was discovered in 2003, in our home state of Washington. She was a six-year-old Holstein and she didn't want to get out of the trailer. It was a busy day at Vern's Slaughterhouse in Lake Moses, and slaughterman Dave Louthan was rushed and "put a hole" in the cow's forehead to keep her from trampling other cattle. Because this cow was able to walk before being shot, she shouldn't have been lumped in with "downers." But the USDA was then paying $10 for samples of downers' brains, so a sample was taken. Before the results of tests on the sample were known, the cow was butchered. Her meat was ground up with twenty thousand pounds of meat from other cows and sold in eight states.[26]

When the USDA lab reports came in positive for mad cow, beef exports plummeted. The nation's beef and cattle industry lost an estimated $3.2 to $4.7 billion.[27] Later it was learned that this cow had been born in Canada and had classic BSE, the sort found in the UK outbreak. She had eaten food contaminated with brain and spinal bits from other cows. It was only by chance that she was shot, mislabeled a downer, and had a slice of brain tissue sent for analysis.

American mad cows number two, three, and four all had an atypical BSE. At least two types of atypical BSE are now known to exist: L-type (aka BASE) and H-type. These occur mainly in animals more than eight years old, animals that are often symptom-free. Several dozen cases of atypical BSEs have now also been detected in other nations: Brazil, Canada, Denmark, France, Germany, Italy, Japan, the Netherlands, Poland, Romania, Switzerland, and Spain.[28]

Mad cow number two in the United States was found in 2005, in Texas. It had L-type BSE. This disease had been discovered only a year earlier in a few cattle in Italy. L-type shows up in a different area of a cow's brain than BSE and does not affect nerves connected to the digestive tract. Experiments have even found the L-type infective agent in the muscle tissue of cows.[29] Steaks are muscle tissue. Researchers don't think the Italian cows got L-type from feed. They speculate it might have been transmitted through the air, been licked up, or have arisen sporadically.[30] Although there are indications that L-type is more virulent than regular BSE in primates and transgenic mice, it's not known if this might also be the case in human primates.[31] No human has yet been known to catch L-type BSE. But mouse lemurs (a nonhuman primate model) have been experimentally infected with L-type BSE from cattle through the oral route (that is, through food). The French scientists running this experiment concluded, "[I]t is imperative to maintain measures that prevent the entry of tissues from cattle possibly infected with the agent of L-BSE into the food chain."[32]

The finding of American mad cow number three, in Alabama in 2006, yielded another surprise: H-type BSE cows can carry a genetic mutation similar to the one that causes inherited CJD in humans. Mad cow number three was a red crossbred beef cow estimated to be ten years old. She had H-type BSE associated with a heritable mutation in her protein gene,[33] and her heifer (which showed no clinical signs of disease at two years of age) inherited the mutation. This was the first time an inherited form of a TSE had been found in an animal.[34] The USDA was not able to track the cow back to her herd of origin.

Scientists at the USDA's Agricultural Research Service estimate that H-type BSE infects fewer than one in two thousand cows.[35] But 93 million cows divided by 2,000 (a worst-case scenario) means we might have nearly 46,500 cows with genetic BSE in the United States. H-type BSE is transmissible by injection into the brains of cows. Experiments are under way to find if, like L-type, it can also be transmitted through the oral route.

The fourth American mad cow was a Holstein, found in April 2012, in Tulare County, California, the nation's top-producing dairy county. The identity of the dairy has never been revealed. Not all cows that die on a farm are tested for BSE. It was a lucky break that this cow was tested. She was found to have L-type BSE. The USDA's Summary Report, issued a few months later, said that the cow was born in 2001. The renderer had picked up seventy-one dead cows from various farms and had taken them to a transfer station where cows meeting criteria for BSE surveillance were given special tags.[36] This cow had apparently been killed at the dairy because she was unable to stand. Tissue samples were tested at the Food Safety Lab at the University of California at Davis. The positive sample was then confirmed as having BSE by the USDA laboratory in Iowa and the World Organization for Animal Health.

We were disappointed to find two falsely reassuring statements on the BSEinfo website. This site is funded by producers under the Beef Checkoff Program (cattlemen fund the site by paying $1 into the fund each time a cow is sold) and overseen by the Cattlemen's Beef Promotion and Research Board and the USDA. The first statement said: "Atypical BSE cannot be transferred from animal to animal." The second said, "You cannot contract the human form of BSE from eating meat such as steaks and roasts."[37] Neither statement has been verified by science. As mentioned above, L-type has been transmitted from cows to lemurs through the oral route, and H-type BSE is heritable. True, eating steaks and roasts is far, far less risky than eating part of the nervous system, and we found no reports of a person catching mad cow disease this way. But contamination can happen during slaughter, and the muscles of some animals can contain prions. An article in the *Proceedings of the National Academy of Sciences* concludes: "Because significant dietary exposure to prions might occur through the consumption of meat, even if it is largely free of neural and lymphatic tissue, a comprehensive effort to map the distribution of prions in the muscle of infected livestock is needed."[38]

In April 2008, the FDA's Department of Health and Human Services tightened its rules about what foods are considered safe for cows and people. The department tried to balance expense and inconvenience to cattle owners and processors with public health safety. Some cow parts have more infectious prions than others, so decisions were based on how much infectious agent is likely to be found in particular parts of cows, and in cows of various ages.

In its introduction to the new rule, the FDA states: "Data . . . indicate that roughly 90 percent of BSE infectivity is contained in the brain and spinal cord, and only about 10 percent of BSE infectivity is present in the retina, dorsal root and trigeminal ganglia, and the distal ileum."

Under the new rules, the following high-risk materials are banned in food for all animals.[39]

- The entire carcass of a cow with mad cow disease.
- The brains and spinal cords of cattle thirty months of age or

older. (Until a cow infected with classic BSE begins showing symptoms, BSE is essentially restricted to its nervous system.)[40]

- The entire carcass of a cow that has not been visually inspected and passed for human consumption—*unless* the cow is under thirty months of age *or* its brain and spinal cord were removed.
- Mechanically separated beef derived from cows over thirty months of age.
- Tallow (fat) derived from other prohibited materials, if it contains over 0.15 percent insoluble impurities.

Removing the brains and spinal cords from dead cows is messy, imprecise work. Splattering might occur, and small parts might be missed. Many commenters on the 2008 regulation said that not all brain and spinal cord *can* be removed.[41] Also, equipment might not be well cleaned, leading to cross-contamination. No one knows how many BSE prions a cow—or you—must eat to become infected. Feeding cows extremely low doses (one milligram) of infected neural tissue has been shown to cause BSE.[42] Given the impossibility of recruiting humans for a clinical trial, even less is known about how much exposure to the BSE agent is needed for a human to develop vCJD.

With regard to human food, cosmetics, and supplements, the FDA bans specified risk material (SRM) from cattle thirty months and older. In addition to the brain and spinal cord, SRM includes the skull, eyes, trigeminal ganglia, most of the vertebral column, tonsils, the distal ileum of the small intestine, and dorsal root ganglia. If a cow can pass a visual inspection and is less than thirty months old, its SRM can go into human food or supplements.[43] But BSE has been found in cows younger than thirty months.[44] So it would seem wise to ban from human dinner plates SRM from an untested cow of any age. And because TSEs are more easily transmitted among members of the same species, it would also make sense to ban SRM from cow food.

Michael Hansen, senior scientist for Consumers Union, testified that the FDA was "clearly bending to the economic concerns of the feed industry at the expense of public health."[45] The FDA counter-argued: "Proposing to prohibit tissues containing approximately 90 percent of BSE infectivity, rather than the full list of SRMs, *provides protection in proportion to the BSE risk in the United States*" (emphasis added).[46] But as explained later, the study on which the FDA's assumptions are based was hopelessly flawed.

Holes in Uncle Sam's Mad Cow Fence

These new food safety laws and what the USDA calls its "interlocking safeguards" to prevent an outbreak of mad cow are a very good change, but still inadequate:

- Although thirty-five million cows are slaughtered annually, only forty thousand are tested for BSE. That's only a bit more than one-tenth of 1 percent and is inadequate. Farmers don't have to report suspicious downer cows for testing, and the USDA forbids citizens (including cow owners) from testing cows at their own expense.
- There is only spotty testing of human brains to find if someone died of vCJD.
- Federal agencies lack the staff, budgets, and sometimes the desire to adequately enforce existing laws and regulations.
- The United States has not yet fully implemented an effective system of tracking cows.

- Legal methods of disposing of dead cows might spread mad cow disease.
- It may be impossible to cut all high-risk material out of a carcass headed for human consumption.
- Cow remains and cow blood can still be fed to chickens, and spilled chicken feed mixed with chicken droppings is fed back to cows. Cow blood plasma, which can contain prions, can still be fed to cows and is sometimes fed to calves in their milk replacer because it's a cheap source of protein.
- Raw cow brain tissue can be legally sold as a human health supplement.[47]

The first six loopholes are covered in more detail below.

After the first mad cow was found in the state of Washington, the USDA launched an eighteen-month study, during which it tested nearly 788,000 cows for BSE. It found only the one mad cow in Texas and the one in Alabama.[48] The USDA concluded that the prevalence of BSE in American cattle was "extremely low," most likely less than one cow per million among adults. Then the USDA promptly reduced the number of cows it tests to about forty thousand a year. This was a mistake, for reasons explained below, because the testing program paid the most attention to those cows the least likely to have BSE.

The USDA's own inspector general declared the 2004–2006 testing process seriously flawed, and called the statistical conclusions the USDA drew from it "unwarranted."[49] Although the USDA did twenty thousand "random" tests at forty slaughterhouses, *it publicly identified those slaughterhouses in advance*, a great service to farmers or ranchers with iffy cows. "Because of the voluntary nature of its program . . . we could not determine how successful APHIS [Animal and Plant Health Inspection Service] was in obtaining a representative proportion of high-risk cattle for testing." The inspector general also questioned the failure to do further testing when there were conflicting test results.

Today the USDA asks farmers to *voluntarily* provide the brains of their downer cows for testing. This can be viewed as an intelligence test for farmers: There is no penalty for not submitting a brain for testing, but if a brain is submitted and it tests positive for mad cow, all hell will break lose. According to veterinary experts who talked to American beef producers privately and were then interviewed by *New Scientist* reporter Debora MacKenzie, "Shoot, shovel, and shut up," is the prevailing attitude.[50]

Creekstone Farms in Arkansas City, Kansas, wanted to test each of the three hundred thousand cattle it slaughters every year for BSE. It estimated that the test would increase the cost of its beef by about 10 cents a pound. Facing boycotts of untested beef in foreign markets, and believing there was consumer demand for tested products in America as well, in 2004 Creekstone invested half a million dollars in a testing lab and hired seven biologists and chemists to run it. (That same year, Consumers Union polled 1,085 adults, 95 percent of whom ate beef, and found that 77 percent of the beef eaters would pay more for beef certified free of mad cow disease.) Japan, which had previously represented 30 percent of Creekstone's market, had stopped importing American beef after the mad cow was discovered in Washington State. By doing its own testing, Creekstone was confident it could win back, or expand, its market share. This is how markets should work—a producer sees an opportunity and invests its own money to address it. If the gamble pays off, then the competitors follow suit.

Meatpacking companies rolled out their heavy artillery against Creekstone's proposal, as did the National Cattlemen's Beef Association. Jan Lyons, president of the Cattlemen, expressed the group's fears: "If testing is allowed at Creekstone . . . we think it would become the international standard and the domestic standard, too."[51] From a consumer's viewpoint, this would be a wonderful outcome. From industry's viewpoint, tests would cut into profit margins. The great unspoken fear was presumably that tests might find many mad cows, and less beef would be sold. When Britain used a test designed to detect abnormal prion protein in cow brain tissues, it discovered BSE in 1,117 healthy-looking cows in its far smaller cow population (America has about nine times as many cows).[52]

Nevertheless, the association's plea landed on receptive ears (the USDA secretary's chief of staff previously had been the chief lobbyist for the association).[53] The USDA flatly prohibited Creekstone from testing its meat for mad cow disease. The ostensible reason was that a "false positive" might undermine consumer confidence in the agency's use of random testing. This isn't a true concern because false positives would always be double-checked, if not triple-checked.

Creekstone sued. The USDA had banned the sale of test kits under an obscure 1913 law, the Virus-Serum-Toxin Act, designed to protect farmers from ineffective medicines. But BSE test kits are not medicine, and the government could hardly challenge their value since it used the very same test kits itself. The federal district court in Washington, D.C., ruled in 2007 that the USDA had exceeded its mandate. The USDA appealed. The D.C. Court of Appeals overturned the district court, permitting the ban on private testing.[54]

The USDA still defends it position: "Since most cattle are slaughtered in the United States at a young age, they are in that period where tests would not be able to detect the disease if present. Testing all slaughter cattle for BSE could produce an exceedingly high rate of false negative test results and offer misleading assurances of the presence or absence of disease." It's true that present quick, inexpensive tests can't pick up an infection from its inception. But not all cows are killed at a young age, and tests can detect BSE two to three months before a cow shows any symptoms.[55]

The Bio-Rad TeSeE ELISA rapid test is the initial test approved by the USDA. The ELISA costs about $20 per carcass and delivers results in a few hours. The likelihood of its failing to find a single case among one thousand samples is said by researchers to be in the range of 1.1 to 3.6 percent. For a while, Japan tested all of their cows for BSE. According to Alberta's Agriculture and Rural Development agency, Japan found an even lower rate for false positives: 1 in 30,000. If a test is "inconclusive," further testing—immunohistochemistry staining and the definitive BSE test, a Western blot test—are done to find if the sample is truly negative. (Tests are done on the obex, a very tiny part of a cow's brain stem, where infectious prions first build up.)

The costs of *not* testing all cows can be high, even when it doesn't lead to a United Kingdom–type epidemic. Early in 2008, visibly sick cows were slaughtered at the Hallmark/Westland Meat Packing Company in Chino, California. Cows that can't walk may well carry BSE or a variety of other diseases. The facility produced ground beef for the school lunch and food assistance program, as well as for burger chains. The Humane Society took a secret video of the cruel, illegal activity that occurred at the packing plant in a single day. Downer cows were given electric shocks, kicked, beaten, dragged using chains, and moved by forklifts to the processing plant. To get them to stand up and walk to slaughter, cows were rammed with forklifts, jabbed in the eyes, blasted with water to simulate drowning, and shocked in the rectum. This was done while five USDA inspectors were on the premises. One of these inspectors was a veterinarian charged with keeping sick cows out of the food supply—someone who should have been notified whenever there was a cow that couldn't walk.[56]

The Humane Society turned the sickening evidence over to local prosecutors, who asked the society to keep the video confidential while they investigated. After a couple of months during which the prosecution made little visible progress, the society gave the video to the USDA and released it to the press.[57] But by then the beef from sick cows had already been commingled with other beef from numerous processors and distributors. Whatever diseases the sick cows carried could have entered the food supply along with their meat. Officially, 143 million pounds of meat products were recalled and destroyed—the largest recall of beef in American history. However, a Costco representative estimated at the time that the real amount exceeded a billion pounds.[58] Despite the massive size of the recall, it may have been futile; nearly all the meat had already been consumed. Since the incubation period for vCJD can be very long, it's too soon to conclude that none of the downers had mad cow and that no people were infected.

Improved tests will soon make possible an even stronger case for universal testing of cows for BSE.

More testing for prion diseases is also needed on people. We don't know how widespread CJD (and possibly vCJD) really are in Americans. Clinical methods to determine whether someone has classic CJD or vCJD allow much room for subjective opinion. Doctors consider such factors as age, duration of the illness, and symptoms. The median age of classic CJD patients, according to the CDC, is sixty-eight. Patients with vCJD have a median age of twenty-eight. The illness usually lasts about four to five months in CJD patients, thirteen to fourteen months in vCJD cases. Dementia and neurologic problems show up early in CJD, but vCJD is characterized by behavioral and psychiatric problems and unpleasant sensations of touch. There is overlap in all these clues.

An MRI brain scan may show patterns of degeneration typical of CJD. But the best way to confirm a diagnosis of BSE, CJD, vCJD, or Alzheimer's is to do a brain biopsy or autopsy.[59] Brain biopsies, which involve removing a sliver of brain from a living person, are inherently risky. Pathologists may shy away from doing autopsies on likely TSE patients; they worry about catching a nasty, fatal disease. Most private insurers, Medicaid, and Medicare won't pay for autopsies, which cost around $3,000 to $5,000. Hospitals lose money on autopsies.

The CDC believes that about 85 percent of CJD cases in the United States are sporadic (for mysterious reasons a prion in a person's brain changes shape and becomes infectious, setting off a chain reaction). The incidence of CJD (adjusted for our growing population) has remained relatively stable since 1979: one death a year for every one million people. This is good, because a rise in numbers would raise the specter of a hidden BSE/vCJD epidemic. Unfortunately, we can't totally rely on the numbers. The CDC monitors death certificates compiled by its National Center for Health Statistics and encourages (but does not require) doctors and medical examiners to report CJD cases. If a certificate says someone died of CJD, and if that person was younger than fifty-five, and if the case was reported, the CDC takes a closer look at the victim's medical records. But only state governments have the authority to mandate the reporting of various diseases.[60] As of 2008, eight states still did not even mandate reporting CJD or vCJD to state health officials. Furthermore, not a single state requires an autopsy when it is suspected that someone died of CJD, vCJD, or Alzheimer's.[61]

Nobel laureate Stanley Prusiner thinks Alzheimer's and other neurodegenerative diseases (amyotrophic lateral sclerosis, Parkinson's) might be prion diseases. He said this twenty years ago and recently repeated his belief.[62] In one study of forty-six patients who had been clinically diagnosed with Alzheimer's, autopsies disclosed that six (13 percent) actually had CJD.[63] Other autopsies of presumed Alzheimer's patients discovered that 3 to 5.5 percent actually had CJD.[64] In 2012, the CDC estimated that 5.4 million Americans have Alzheimer's. If, say, 4 percent of them actually have CJD, that would equal 216,000 cases of CJD—far, far higher than the 314 deaths per year that would be expected among the U.S. population. If we're missing huge numbers of cases of CJD, might we also be missing cases of vCJD among older persons? Bear in mind that the incubation period for vCJD can last for decades.

The CDC's "CJD Surveillance" website states: "If current control measures intended to protect public and animal health are well enforced, the cattle epidemic should be largely under control and any remaining risk to humans through beef consumption should be very small."[65] Under any conceivable definition of "well enforced," both the USDA and the FDA fail the CDC standard.

The USDA's Food Safety and Inspection Service is responsible for inspecting cows before they are slaughtered. These "inspections" are typically just quick visual once-overs. Regulations require an inspector to be present for slaughtering to occur. But the FSIS budget is kept so low that a single inspector is often responsible for as many as two dozen facilities spread over a large geographical area. A survey in 2007 conducted by the food inspection unions found that *three-quarters of inspectors didn't make their required daily visits.* And even when they were on site, lines moved too quickly for them to do their job well.[66]

In Republican and Democratic administrations alike, both the USDA and the FDA often seem more anxious to protect industry interests than the public. Here's a classic example. In 2005, when USDA field scientists produced inconclusive results as to whether a tested cow had mad cow disease, they recommended it be retested. Top officials ordered them not to perform the additional tests because they feared a positive finding would undermine confidence in the agency's testing procedures. In one of those marvelous, subversive processes that sometimes seem to be the last best hope of the federal bureaucracy, the material was preserved and the department's inspector general was alerted. The inspector general protested, the sample was sent to England for retesting, and mad cow was found. It was the second case of mad cow in the United States, and like the first case it very nearly wasn't discovered.[67]

Cows are an ambulatory form of property. Ever since cows and people domesticated each other, people have needed some way to pick their cow out from the herd, some proof of ownership. In the American West, cattle brands are icons, avatars, a cattleman's coat of arms. Branding cows is a cherished part of the rough lifestyle. A brand is hard to remove and easy to see.[68] But brands disappear along with a cow's hide.

After a cow's death, its remains travel far. Cow products are shipped around the world. No one paid much attention to this postmortem perambulation until mad cow disease popped up in Great Britain. Like us, the Brits have incorporated cows and cow products into every nook and cranny of their lives. When mad cow disease emerged, they wondered whether prions might survive their transformation into things other than food. But when they tried to track down where cow parts went, the task proved to be overwhelming. According to the 2000 British "B.S.E. Inquiry Report," "It has been said, and not altogether facetiously, that the only industry in which some part of the cow is not used is concrete production."[69] They were wrong: In some countries cow products have even found their way into cement blocks.

Here in the States, not all farmers and ranchers want their cows tracked. They may object to the cost, the invasion of their privacy, or the imposition of yet another disadvantage on the small farmer. The public, however, has good reason to want cows tracked. Learning where an outbreak originated makes it possible to tailor a recall to fit the need, to punish the sloppy, and to reward the responsible. Mad cows aren't the only things that need tracking; *E. coli* and other infections or toxins need to be quickly traced to their origins.

The United Kingdom, the European Union, and Australia already track their cows. A new law in the United States, the Animal Identification Management System, was issued in January 2013. Every cow born in this country is to get a unique fifteen-character ID number, and when a cow crosses a state line a certificate of veterinary inspection must accompany it. When a cow is slaughtered, its official ID is to be collected at the slaughterhouse, put in a clear plastic bag, and fastened to the carcass until after a postmortem exam is complete.[70] After meat leaves the slaughterhouse, tracking ends.

The new plan is mandatory, but there are exemptions. Amid budget difficulties, it is uncertain that Congress will appropriate money for implementation. It's possible to trace meat all the way from the slaughterhouse to shiny shrink-wrapped packages, but only those who raise specialty (for example, organic or grass-finished) cattle are likely to even consider doing this.

Entrepreneurs see tracking requirements as a great business opportunity. A quick online search revealed a dazzling array of cow earrings for sale, including official USDA tags. There is even a new type of identifying tattoo for cows: Needles inject radio-frequency identification (RFID) ink, which can be colored or invisible. The maker thoughtfully notes that the ink is harmless for humans to ingest, and that by chewing the meat you'll make it impossible for the government (or vegetarians) to track you down.[71]

• • •

Mad cow disease has made it challenging to dispose of dead cows. Most unprocessed dead cow parts (including brain and spinal cord material from cows over thirty months old) are put in landfills. In sparsely populated areas dead cows are buried, burned, incinerated, or simply abandoned.[72] It's legal in many states and localities to bury cows.[73] But it can take a quarter-century for a buried cow carcass to fully decompose, and as it decays it may release harmful chemicals and infectious agents into the soil. Infectious BSE prions can bind to clay and minerals and survive for years in soil. When this happens, they might transmit the disease to other species.[74]

There is no evidence yet that cows or people have become infected with BSE from eating crops grown in infected soil or drinking infected water. But some scientists worry about this and hypothesize that prions in cow feces and urine that is spread on fields, put into landfills, washed into sewers from slaughterhouses, and so forth, can get into the groundwater and streams and remain capable of spreading infection even after going through wastewater treatment systems.[75]

Composting cows is also legal in many states. It needs to be done where it won't contaminate groundwater or cause too much stink. But cows contain more than just cow—we pump them full of antibiotics, growth promoters, vaccines—and those chemicals can survive composting. Heavy metals, bacterial spores, and infectious prions can also survive composting.

A popular choice when a cow has died of a disease is to burn it. To neutralize prions, the CDC's biosafety guide recommends incineration at 1,832 degrees Fahrenheit.[76]

It's also possible to dissolve a dead cow. A container for alkaline hydrolysis tissue digestion can be made large enough to fit a cow into (Dumpster size), along with a lot of potassium hydroxide (lye). Digesters can dissolve a cow into a sterile, syruplike fluid that can be composted or dumped into a sanitary sewer (if temperature, pH, and biological oxygen demand are monitored). Digesters reduce a dead cow's volume and weight by up to 97 percent and don't need to release anything noxious into the air.[77] Odors are minor. Prions and other infectious agents are destroyed. You can see photos of a digester in action on the Internet, turning cows into a soapy, coffee-colored liquid.[78]

Many leftovers are rendered. Rendering means grinding up bloody, fatty animal waste, cooking it, removing the fat, and drying the remaining sludge into pellets or powder. It's a revolting, messy, stinky, but essential business. The pellets and powder are used by other industries to make animal feed, fertilizers, alternative fuels, and ingredients for personal care products, paints, gelatins, and plastics. Rendering completely destroys all known infectious agents . . . except prions.[79]

James McWilliams, an environmental historian at Texas State, did a piece for the September 2010 online version of the *Atlantic*, "The Deadstock Dilemma: Our Toxic Meat Waste." Because of regulations to prevent the spread of mad cow disease, the cost of rendering has gone up as much as fivefold, McWilliams says, and these costs are passed along.[80] Small local operations can't afford necessary updates, and they have folded or been bought out by large operations. We did some research of our own and found that the rendering industry is rapidly consolidating into yet another Big Ag bloc. For example, Darling International has acquired Boca Industries, Sanimax USA, and Nebraska By-Products. In March 2011, Darling merged with Griffin Industries, resulting in a huge rendering conglomerate with 136 facilities in 42 states.[81]

Mad cows have faded from American minds. Complacency in the United States today is comparable to that in Great Britain in 1993. In January 2011, an article in *New Scientist* trumpeted that mad cow disease was now almost extinct, its eradication a triumph for science.[82] In June 2013, a syndicated columnist commented in a *Seattle Times* op-ed, "Seeing as there's no reported case of anyone's getting sick, much less dying, from an American mad cow, perhaps the monitoring is adequate." This statement reflects a major lack of understanding of the disease, because one can't just "get sick" from mad cow and not die.

The holes in Uncle Sam's mad cow fence need to be plugged now. The decades-long BSE incubation period could obscure the presence of an epidemic in humans until it was widespread. In the meantime, you can do what we do: Buy organic, grass-finished beef, or buy bison; and if you crave a cow burger, grind your own meat.

Chapter Eight

WHY BUY ORGANIC BEEF AND DAIRY?

Health-conscious Americans are buying more and more organic food and are willing to pay a premium for it. Organic food sales grew from about $1 billion in 1990 to over $32 billion in 2012.[1] Naturally, corporate agribusiness recognized an opportunity and played a major role in this industry's growth. Global giants like General Foods, Kraft, and Groupe Danone shouldered into territory once occupied solely by small local farms, artisanal specialists, health food stores, farmers' markets, and community-supported agriculture groups (CSAs). They weren't welcomed with open arms.

Small organic farmers and ranchers are appalled that the organic brand has been tainted by profit-maximizing outfits with little heart in the game. Multinational firms scoff at the organic purists as marginal zealots blind to the need to feed seven billion people.

When General Mills began marketing a Cheerios brand as "not made with genetically modified ingredients" in 2014, organic enthusiasts erupted with indignation. They pointed out that no one has yet even developed genetically modified oats, and that General Mills's biggest change was adding the new label. (The actual change was to stop using some starch and sweetener ingredients made from a type of corn that is classified as a genetically modified organism, or GMO.) Don't get the idea that General Mills thinks genetically modified food is a bad thing. The company's vice president for global communications wrote a blog post emphasizing that the product was basically unchanged and that General Mills was still a full-throated advocate for GMO crops.[2] The company also weighs in with its wallet, supporting efforts to defeat state-level legislation that would require GMO foods to be labeled as such.[3]

Labels matter. Customers these days almost never know the farmers who produce their food, and wouldn't have the know-how to evaluate farm practices even if they personally visited farms. A trustworthy certification that a product meets rigorous standards is valuable information. To fill this need, Senator Patrick Leahy, who then chaired the Senate Agricultural Committee, sponsored the National Organic Program. Foods certified as 100 percent organic can now bear the "USDA Organic" seal, plus the rarely seen label "100% organic." If they are 95 percent organic, and the other 5 percent of ingredients can't be obtained in organic form, they can still legally carry the "USDA Organic" seal.

But what's "organic"? The National Organic Standards Board (NOSB) determines this. On paper, the group is broadly representative: Four places are reserved for farmers, three for environmentalists, three for consumer advocates, two for processors, one for a retailer, one for a scientist and one for a USDA certifying agent. But the Secretary of Agriculture, who is required to consider the interests of agribusiness as well as those of the public, appoints the members. The first draft of organic standards produced by the Board allowed irradiated and genetically modified food to be called organic, as well as food grown on sewage sludge laced with industrial waste. The USDA was flooded with letters and comments opposing these "big three" enemies of the grassroots organic community, and the agency agreed to exclude such food. Synthetic substances are usually prohibited unless (like certain vaccines) they are specifically permitted, and naturally occurring substances are usually permitted unless (like arsenic) they are specifically prohibited.[4]

The description of occupational categories from which NOSB members must be drawn is distressingly vague. For example, the single slot available for a "scientist" could be filled by someone who works for a pesticide manufacturer, and the assignment of nominees to reserved places seems almost random.[5] On only three occasions since 1992 has an environmental slot been filled by someone who actually works for an environmental group. Organic farmers, environmentalists, and consumer advocates have generally been underrepresented and their slots filled by those more willing to be flexible—the USDA would say pragmatic—about what qualifies as organic.

Trying to define organic food as a "movement" is tricky. Most people who became organic farmers in the 1960s and 1970s didn't think of themselves as part of an organic crusade. They were just part of The Movement—against the Vietnam War and the nuclear arms race, supportive of civil rights, feminism, and the environment. This remains part of their identity. They believe in eating local, seasonal food. They usually farm on a small scale.

Today the Organic Consumers Association, to which many of them belong, campaigns "for health, justice, and sustainability" and supports energy independence and renewable energy, universal health care, and fair trade as opposed to free trade. It ardently opposes GMO foods and supports labeling them as such. Although some of its goals may differ from those of even enlightened industrial organic interests, the Organic Consumers Association works closely with the giants whenever it can.

Then there is John Mackey, co-CEO of Whole Foods, who has arguably done more than anyone else to bring organic produce to Americans. He is emphatically not part of the mildly leftish movement. An articulate libertarian and admirer of Ayn Rand, he decried the Affordable Care Act ("Obamacare") on NPR as "fascism" and expressed his belief that climate change is natural and not such a bad thing.[6] In contrast, another international company in the organic market, the Rodale publishing empire, remains committed to the sweeping principles of its founder, John Irving Rodale. It blends organic farming with advice on diets and health, and has expanded into magazines like *Bicycling*, *Runner's World*, and *Prevention*. Rodale published Al Gore's best-selling climate book, *An Inconvenient Truth*.

A great many smaller organic companies have been snapped up by big corporations like Hain Celestial, General Mills, Kellogg, and Coca-Cola. The organic product usually continues to be marketed under its old corporate name, so buyers might not be aware of the change in ownership. On its website, the Cornucopia Institute posts a diagram that makes it easy to see which big corporations control which brands.[7] Major grocery chains and retail outlets have also launched their own organic brands.

Among nationwide organic dairies with excellent ratings from Cornucopia are Organic Valley and Whole Foods 365.[8] George Siemon, Organic Valley's longtime CEO, describes the huge co-op as "a social experiment disguised as a business."[9]

New tensions constantly arise. One is over imported organic food. Private companies and government agencies around the world are allowed to certify farms. American organic farmers with a dream of healthy, sustainable, seasonable local produce are aghast at the idea of cheap "USDA Organic" produce flown in from abroad in hard plastic containers and sold at supermarkets and bulk stores in the middle of winter. They believe this casual labeling misleads consumers. The supermarkets agree that such imported food undermines some of the goals of organic enthusiasts but argue that it is still better than nonorganic food.

In 2006, Wal-Mart, the biggest grocery retailer in the States, announced a push to sell more organic food in its stores to attract more upscale buyers, and to sell it for only about 10 percent more than conventional equivalents. By 2007 more people shopping for organic food did so at Walmart stores than at any other place. Other big chains have followed Wal-Mart's lead, often with their own brands. The Organic Consumers Association had doubts about this trend. They noted that the demand for organic food already exceeded the supply, and worried that Wal-Mart would start importing dubious organic food.[10] By 2012, the company was indeed having trouble rustling up enough organic produce, so its organic emphasis became focused on products like baby food and milk, not fresh fruit and vegetables. Wal-Mart also promised in 2010 to buy food from local farmers, by which the company meant food grown in the same state as a retail outlet.[11] In giant-sized California with its giant CAFOs, "local" loses its meaning.

Our Beef with "Sustainable Beef"

On January 7, 2014, Mcdonald's announced that it would begin purchasing "verified sustainable beef" in 2016. With a $95 billion market capitalization, McDonald's is the largest burger chain on the planet and its announcement made news. However, the company gave no indication how much beef it would buy in 2016, merely that it would buy some. "Verified" is an ambiguous word here but clearly doesn't mean "certified by an independent third party." And "sustainable" is as slippery as "natural."[12] In fact, *Bloomberg Businessweek* titled its story about the announcement, "Fleshing Out the Incredibly Vague Concept of 'Sustainable Beef.'"[13]

McDonald's has taken a number of important environmental steps for which it has received little credit. It already uses only seafood that has been certified by the independent Marine Stewardship

Council and only palm oil that has been certified by the Roundtable on Sustainable Palm Oil. A full 25 percent of McDonald's coffee worldwide (and 100 percent of its espresso in the United States) has been certified by both the Rainforest Alliance and by Fair Trade.

Sustainable beef, however, is dramatically more complicated and also a much bigger deal for McDonald's. The company has five beef sandwiches—Big Mac, Quarter Pounder, Quarter Pounder with Cheese, Cheeseburger, and Double Cheeseburger—each generating more than a billion dollars in annual sales. Its influence on the rest of the industry is enormous. So when it comes to the nation's cows, its actions on beef matter.

The beef supply chain, which includes corn famers, cattle ranches, feedlots, dairies, processors, and packing plants, is long and complex. An estimated four hundred thousand farms raise cows that wind up as beef at McDonald's. The typical hamburger (in the United States) often contains beef not just from different farms in different regions but also from different countries on different continents. Tracing beef from a sick cow to hamburger today is impossible, as is tracing beef back to an organic farm or a huge feedlot. Determining how much water, fertilizer, and herbicide a corn farmer used to produce the grain a feedlot used to fatten a cow is laughably impossible. So journalists were inclined to give McDonald's the benefit of the doubt when it announced a nebulous goal without a program to achieve it.

The Global Roundtable for Sustainable Beef—the vehicle that McDonald's embraced for this campaign—issued its first draft principles and criteria for public comment in March 2014. These were broadly aspirational. The principles sound good, although most are so imprecise that it's hard to imagine who could disagree with them. The goals do include such things as maintaining healthy animals and paying at least the minimum wage. (Aspiring to obey the law isn't much of a stretch, however.). The draft mentions vague concepts such as "information-sharing platforms" and "best practices" and "linking value-chain partners." But it does not contain the words "grass-finished," "antibiotic-free," "organic," or "family farm."

The drafters note, correctly, that they are trying to build consensus in a fragmented industry—the Roundtable includes Wal-Mart, Cargill, JBS, and a variety of ranchers, processors, restaurants, retailers, environmentalists, and food activists. This document, they say, is just a first tentative step down a long road.

Skeptics, who are legion, view McDonald's sustainable beef gambit as corporate greenwashing, an effort to score consumer points with words instead of deeds. McDonald's, on the other hand, sees the undertaking as a transparent effort to establish a bold goal even before the company has a clue how to achieve it or exactly what that goal is.

Bob Langert, McDonald's vice president for global sustainability, believes that the company is merely responding to its customers. In an interview with Joel Makower, executive editor of greenbiz.com, Langert noted, "From the research we do, customers really care about where their food comes from." He ranked nutrition and obesity as top concerns but placed food-sourcing issues like animal welfare and antibiotics close behind. "We think they are going to vote with their feet more and more on this issue," he said. Langert's boss, J. C. Gonzalez-Mendez, was equally emphatic about being customer-driven: "Sixty-nine million people visit us every single day. They like their food and they want to know more about where it comes from and how animals are being treated." It is testimony to the uneasiness Americans feel about the way beef is now produced that this huge, successful, consumer-driven company feels pressured to do *something* about its beef.

So what's our beef with the Global Roundtable for Sustainable Beef? The Roundtable is all inclusive: Cattlemen, cattle processors, retailers, fast food giants, international environmental groups, and others who disagree on many things are all represented. No one can dictate terms to the others; all must voluntarily concur. This gives an absolute veto to any economic interest whose ox might be gored (as it were). Change isn't ordinarily initiated with the full concurrence of all those whose worlds would be turned upside down. And members share three fundamental assumptions that are incompatible with any reasonable definition of sustainability: America will produce more and more beef as the century unrolls, the price of beef will remain low, and the basic models of production won't change in any fundamental way.

The McDonald's sustainable beef initiative is justified in the company's eyes as a method to foster growth, primarily through improving the efficiency of the production process. "This is totally generated out of us devising a strategy, wanting to grow our business, and seeing the opportunity to be a leader in this area and derive some business benefit as well," Langert says.

McDonald's is in the business of selling hamburgers. It has attempted to sell many other things, with lackluster results. To meet the expectations of its shareholders, it must sell more beef every year. For reasons rooted in the biological carrying capacity of our land, that cannot be done sustainably.

McDonald's isn't the restaurant of choice for the 1 percent. It provides food for the rest of us, and it is proud of its affordability. "Sustainability should not be some sort of niche, premium, extra-cost endeavor that's for a very narrow segment of society that has enough means and wealth to purchase sustainability," says Langert. "The fact is, sustainability belongs in the masses."

It's hard to argue with that without risk of sounding callous. But cheap beef is the product of huge hidden subsidies that harm people, the land, and cows. Take away the huge subsidies for corn, the generous provision of nearly free water, begin to treat and regulate cow sewage as we do human sewage, assign penalties for dead zones downstream, regulate greenhouse gases, and the consumer will begin to see the real price of his hamburger. Truly sustainable beef production is production that assures the land will still be able to support as many cows a thousand years from now as it does today.

The same amount of money can purchase a lot of cheap, less healthy beef from miserable cows, or a smaller amount of more expensive, healthier, organic, grass-finished, humanely raised beef. Either diet will amply meet the consumer's need for protein. But the fundamental business proposition at McDonald's is that lots of cheap beef is better than a smaller amount of more expensive beef. The company appears to be worried that if it defines "sustainability," which its customers want, in a way that is honest—that is, if it acknowledges that sustainability is inherently more expensive—those customers will go next door to Burger King. And unless Burger King and Wendy's and Arby's all use the same honest definition of sustainable, McDonald's may be right.

"If you are a trader and your margin is one point five to three percent and the cost of certification is one percent, it is a huge part of your profits. That's why the business case is so important. We don't think that a premium is on the table for any of this stuff." So says the World Wildlife Fund's lead negotiator at the Global Roundtable. Getting to anything a biologist might recognize as sustainability is likely to cost far more than 1 percent. If 1 percent is unthinkable, even in the eyes of the top environmental negotiators, nothing very important will change.

Many talented, committed people are putting time, money, and effort into the Global Roundtable. Years down the road, beef production will probably be marginally better as a consequence. But the changes will be limited to highly visible shifts that can be easily accommodated within the status quo.

At the risk of overgeneralizing, this captures the essence of why large environmental groups have achieved so little in their efforts to collaborate with business interests. The major green groups in the "collaboration niche" try, naturally, to maximize their impact. The impact of nudging McDonald's or Proctor & Gamble or Duke Energy even a little bit toward sustainability is vastly greater than the impact of persuading a local mom-and-pop restaurant to be deeply sustainable. Unless, of course, the radically green local restaurant turns out to be Chipotle and it starts growing six or seven times as fast as McDonald's by making a sincere, deep green commitment to sustainability.

A great many firms, mostly new and headed by younger people, are springing up—the Chipotles, Burgervilles, Bareburgers—and are trying to do things right. Readers of books like this will spend their dollars at these places. Of course, virtue doesn't, by itself, produce success. Success requires finance, procurement, marketing, training, and great recipes. But some of the businesses that focus on high-quality organic, grass-finished beef are going to be astonishingly successful. And as they capture an ever-greater share of the market, they will produce far more change at the beef and burger giants than anyone in the Global Roundtable currently thinks possible. In the words of Samuel Johnson, "When a man knows he is to be hanged in a fortnight, it concentrates his mind wonderfully."[14]

Does Organic Beef and Dairy Make Sense?

Critics of organic food argue that there's not enough arable land in the United States to feed all of us organically and that organic food costs too much for most people to afford. If either claim were true, it would be compelling. However, both are false.

The argument that all good farmland is already in use, and that not as much food can be produced per acre if organic methods are used, is bogus because most good farmland is now being used to feed cows and other livestock and to make ethanol. If we plant more diverse crops low on the food chain, we have ample land to feed everyone organically. Furthermore, fifty thousand square miles of irrigated land in the United States are used to grow lawns—grass that isn't eaten. This "crop" has been estimated to be three times the acreage of irrigated land used to grow corn, and it's thirsty—sucking up as much as 200 gallons of good water, per person, every day.[15]

Brazil has the largest population of commercial cows in the world—some two hundred million head. Ninety-six percent of them eat nothing but grass their entire lives (after weaning). Before settlers exterminated the bison population on the American plains, an estimated fifty million of these one-ton, high-energy animals roamed the land eating nothing but grass.[16] Maintaining forty-five to fifty million cows on grass isn't a technical challenge; it's a matter of willingness to change the status quo.

The argument about cost is also specious, because by eating lower on the food chain most Americans could afford organic food. Recall that poorer Americans tend to eat the most beef. Although underprivileged people are reluctant to give up beef—a status symbol to many—food status tends to follow the choices of the upper classes, which are dramatically reducing their beef consumption. As to affordability, the average American household spends less than 10 percent of its disposable income on food—a world record, of sorts—and beef is its most expensive protein source.

Proponents of organic food argue that organic food production causes far fewer environmental problems than non-organic food, is safer for farmworkers, and is healthier for everyone. Furthermore, they say, when you buy organic products you are often supporting sustainability: family-sized farms whose owners grow their own feed for cows without using pesticides and herbicides, who put a lot of work into enriching and conserving their farm's soil, and who have cows that don't contribute to antibiotic resistance. Such farms might even grow the fuel their tractors use, and they support healthy rural communities.

The harm pesticides do to farmworkers is well documented.[17] Not only do they play havoc with hormones, but they are often carcinogenic and toxic to the nervous system. Pesticides can affect not only adult workers but also their non-farmworking children and grandchildren.

The real fight between the organic and feedlot camps is whether organic food is more nutritious. Why, you might wonder, hasn't science delivered a knockout blow to one side or the other? Nutrition research is exceptionally complex. Ideally, when comparing organic dairy and beef to conventional dairy and beef, researchers should consider the breed, sex, age, and health of the cows; the types of forage available in a pasture; whether forage is at its peak levels of nutrients when eaten; where dairy cows are in their lactation cycles; the length of time cows spent in a pasture; what food cows are given when not pastured; what chemicals they have been exposed to in their environment; what supplements they've taken; the season in which the food was harvested; how hot or cold the weather was; whether the cows were stressed; the average milk yield of the herd; and even the altitude of the pasture. It's also relevant whether beef was tenderized or milk pasteurized, how the product was transported and stored, and how fresh the product was.

It would be desirable to further control for many human variables: the age and gender of the people studied; their ages at exposure (in utero, as children, or as adults); the length, intensity, and timing of their exposure to the food in question; the rest of their diet; genetics and epigenetics; their occupational and household exposure to pesticides, endocrine disruptors, and toxins; whether they smoke; and on and on. Ideally, a researcher would start back at least one generation and learn about the parents and what they were exposed to, and then trace subjects from conception to old age (real-world studies often last just a few months).

On top of this, there's the design of a study: how large it was; whether there were controls; if it was double-blind, prospective, or retrospective (the latter being more dependent on humans' famously fallible memories); and what laboratory techniques and equipment were used.

These lists are far from complete, and of course we found no research on beef and dairy that accounted for all relevant factors. To do so might be impossible. So the results of studies vary. That doesn't mean all studies are worthless. But read them critically.

Food from Pastured Cows Might Be More Nutritious

A 2012 meta-analysis published in the *Journal of the Science of Food and Agriculture* found that organic dairy products contain more of some nutrients that are possibly important: protein, alpha-linoleic acid, total omega-3 fatty acids, certain isomers of conjugated linoleic acid, and *trans*-11 vaccenic acid.[18] We say "possibly" because some studies are conflicting and there isn't clinical-trial proof.

You've probably never gone to bed wondering, "Did I get enough conjugated linoleic acid today?" But maybe you should have (chapter 5). Milk, cheese, and beef are among the best sources of CLA. Forage-fed beef has twice as much CLA as grain-finished beef.[19] Pasture-fed cows' milk consistently has far more CLA than milk from dairy cows on feedlots,[20] and cheese from grass-fed cows has four times as much CLA as cheese from grain-fed cows.[21] It's not surprising that pasture-fed cows have more CLA. Cows are what they eat. This chain of causation continues into humans: When nursing mothers switch to organic meat and dairy products, their breast milk contains more rumenic acid (the main CLA).[22]

The amazing biota in cows' rumens, along with activity in their mammary glands, make CLA out of pasture plants. Eating corn, on the other hand, fills a cow's rumen with fermentation acids that aren't well absorbed or turned into nutrients like CLA that the cow needs. The acidic environment also lets harmful bacteria thrive. (Unlike human stomachs, cows' rumens are normally only weakly acidic.)

Grass-fed beef is higher in vitamins B and E than grain-fed beef, and contains more calcium, magnesium, potassium, lutein, and zeaxanthin.[23] Vitamin B_2 (riboflavin) is particularly important for pregnant and lactating women and for growing children, and milk and milk drinks are Americans' most common source of B_2. B vitamins are all water-soluble and aren't stored in our bodies, so we need to frequently eat foods that are rich in them.

Another benefit is that beef has less total fat when it comes from organically raised and/or grass-finished cows. Milk, on the other hand, seems to have about the same amount of total fat content whether cows were raised in a pasture or conventionally.[24]

Some types of fat are crucial to our health. Both milk and meat from grass-fed cows are likely to contain more of these than does meat[25] or milk[26] from grain-fed cows. Among these good fats are omega-3 fatty acids. Americans consume too little omega-3 and too much omega-6. This imbalance makes us more susceptible to cardiovascular problems, cancer, autoimmune diseases, and inflammation. In infants, omega-3 fatty acids are needed to grow brain cells.[27]

Two of the most important omega-3 fatty acids are eicosapentaenoic acid, or EPA, and docosahexaenoic acid, or DHA, but one out of five Americans has no detectable EPA or DHA in her or his blood.[28] The FDA allows packaged food to bear a label stating: "Supportive but not conclusive research shows that consumption of EPA and DHA omega-3 fatty acids may reduce the risk of coronary heart disease."[29] Canada's Food Inspection Agency's label states: "DHA, an omega-3 fatty acid, supports the normal development of the brain, eyes and nerves primarily in children under two years of age."[30] Even organic beef and milk are not the richest sources of these fatty acids: higher concentrations are found in algae and cold-water oily fish like salmon, sardines, and herring. But comparing cows with herring is comparing apples with nori; if you choose to eat beef or dairy, choose the type richest in omega-3s.

The fattier the beef, the more likely you can cut it with a fork. This fact led the USDA to create a labeling system for beef (Prime, Choice, Select, Standard) that is purely a reflection of its saturated fat content, not its healthiness. In fact, the higher the USDA rating, the less healthy the beef.

Because grass-finished beef contains less saturated fat, its leanness may make it less tender.[31] But it is extremely difficult to judge tenderness until you bite into the cooked cut. Factors other than fat can have a major influence. If a cow is stressed shortly before or during slaughtering, its meat will be less tender.[32] Beef from younger cows is usually more tender. Aging and cooking methods can also tenderize beef.

Interest in grass-finished beef is growing, and innovative ranchers are trying to breed cows like the Murray Grey and Red Angus that offer optimum deliciousness. An article in the *New York Times* by Kathryn Shattuck describes two such ranchers and how they are inspiring a culture of *terroir* for beef. *Terroir* is French for "land."

The word is usually used to describe the qualities of fine wine derived from the land on which the grapes are grown.[33] Beef connoisseurs might someday pride themselves on being able to bite into a steak and announce, "Ah! A Pinzgauer, probably pastured in Idaho, where it grazed on smooth brome, clover . . . and with a high note of alfalfa?"

Don't be fooled by the word "natural" on a cow product. It's meaningless on a dairy product (other than to concede, perhaps, that the product is not supernatural). On a package of beef, "natural" means only that the meat contains no deliberately added artificial ingredients or colors.

"USDA Organic" and "grass-fed" labels are not interchangeable. "Grass-fed" is a voluntary USDA certification label available only to meat producers. "Grass-fed" animals can never be fed grain or grain by-products after weaning and must have continuous access to pasture during the growing season.[34] Nonorganic grass-fed animals may be given hormones and antibiotics. On the other hand, USDA Organic animals can be fed organically grown supplemental grain.

Kick the Toxin Habit

The USDA Pesticide Data Program has found ten pesticides in nonorganic beef. Six of the pesticides are known or probable carcinogens and nine are suspected hormone disruptors. In milk, they have found twelve pesticides. Five are known or probable carcinogens, eight are possible hormone disruptors, four are neurotoxins, and another four have developmental or reproductive effects.[35]

Pesticides get into nonorganic cows through their food. Children exposed to organophosphate pesticides at critical times in their development tend to have lower IQs, lower birth weights, and a higher risk of attention deficit hyperactivity disorder (ADHD).[36] When children are fed organic food, their levels of toxic organophosphates drop rapidly.[37] But the effects of previous exposure might be permanent. *Consumer Reports* concluded: "It's worth it to buy organic versions of the foods that are likely to have the highest levels of pesticides when grown conventionally, as well as organic poultry and milk, to reduce exposure to antibiotics. Those choices are especially important for pregnant women and children."[38] The American Academy of Pediatrics also urges parents not to expose their kids to pesticides. The academy warns that diet often accounts for most of a child's exposure, and that organic diets expose consumers to fewer pesticides associated with human disease.[39]

Nearly half of all babies born in the United States get food from the federal Supplemental Nutrition Program for Woman, Infants, and Children (WIC) during their first year of life. WIC gives states grants to provide nutritious supplemental foods for low-income pregnant or breast-feeding women, and for infants and children up to age five.[40] Participants get checks, electronic benefit cards, or vouchers to buy specified foods from retailers. Milk and cheese are approved foods.

Outrageously, the USDA allows states to list organic foods as "not approved." The rationale is that organic foods may be more expensive. Among the foods banned for WIC in our home state of Washington are organic cheese and milk. Washington's list is fairly typical of other states' lists of forbidden food.

Babies are usually given whole milk. As well as benefiting from greatly reduced pesticide contamination in organic milk, an infant will benefit from the additional omega-3 fatty acids found in organic milk. The USDA's decision to let states ban organic foods from WIC programs is deplorable—and the fact that so many states take advantage of this loophole is unconscionable.

Buying organic food isn't a rock-solid guarantee that it contains nothing harmful. When the USDA Organic label appears on food, it means that the *process* under which the food was grown has been certified—not the food itself. Organic food critics correctly point out that even organic farming isn't completely free of pesticides or toxins.[41] Unfortunately, conventional pesticides can be detected at low levels in some organic foods, usually because of past contamination of the soil or "drift" from neighboring conventional farms.[42]

Organic dairy farmers have found substitutes for manufactured chemicals. They may use peppermint oil and balm or garlic tincture to treat mastitis, for example. For footwart some use copper sulfate and vinegar or tea tree oil. The USDA allows over twenty naturally occurring pesticides to be used on organic farms, and some of them (such as copper sulfate) might be as bad for people and the environment as some synthetic pesticides. The possible dangers of "natural" pesticides are just as poorly studied as the dangers of conventional pesticides. Natural pesticides are often chemicals that evolved in plants to protect them from insects or animals. The chemicals may also protect the food from people; "natural" doesn't mean "safe."

Rotenone is a good example of what was until recently a bad, but allowed, organic pesticide. A naturally occurring substance found in jicama and some other plants, it was widely used on organic crops before 2012. Rotenone works by interfering with the electron transport chain in mitochondria, those tiny powerhouses inside every cell that make life possible. Rotenone kills caterpillars, worms, beetles and other arthropods, beneficial insects, birds, and fish. It was long a common treatment for head lice on schoolchildren. People exposed to rotenone are more than twice as likely to develop Parkinson's disease.[43] It is now being phased out in the United States, except to combat invasive fish.

It would make a lot more sense to allow or disallow chemicals based on the danger they pose to humans and the environment rather than on whether they occur naturally. The manner in which a chemical is used (the amount, the timing of applications) also has a huge impact on its safety. Some synthetic pesticides are safer than some natural ones while also being more effective. For example, a Canadian study of ways to control soybean aphids compared natural organic pesticides with some new synthetic ones. It found that the natural pesticides were less effective, had greater negative effects on the aphids' natural predators, and had wider negative environmental impact than the targeted, selective synthetic pesticides.[44]

On the other hand, most naturally occurring pesticides, but not all, tend to break down in the environment more rapidly than synthetic chemicals, and pesticide residues are much greater on conventionally grown crops.[45] Many synthetic insecticides and herbicides contain organophosphates, which scientists know are highly toxic to humans and wildlife. People sometimes swallow organophosphates to commit suicide. Organophosphates also have insidious effects not immediately apparent. Consider this: Your own DNA and RNA are organophosphates. So is nerve gas. In adults, exposure to organophosphates might increase the risk of dementia and a great many other health problems.[46]

Roundup (glyphosate) is an extremely popular synthetic weed-killing chemical that is absorbed through foliage. After Monsanto developed genetically modified crops that were not harmed by it, Roundup's popularity soared, because Roundup could be sprayed on those crops and kill only weeds. Today, 85 percent of the corn grown in the United States, and 93 percent of the soy, is genetically modified to be herbicide-resistant.[47] There are also Roundup Ready seeds for alfalfa, canola, cotton, and sugar beets.

Roundup is an organophosphate, but it targets a metabolic system in plants that humans lack, so by itself it is not acutely toxic to humans. Glyphosate's potential to disrupt endocrines and mutate genes is disputed; but the EPA decided to include glyphosate in its Endocrine Disruptor Screening Program. It's the other chemicals that are routinely added to Roundup that are thought by many scientists to be considerably more dangerous than glyphosate itself.

The first few years after Roundup Ready soy and corn crops came into use, fields required less herbicide. But weeds quickly evolved to also become resistant to glyphosate. A peer-reviewed study by Washington State University professor Charles Benbrook found that the use of genetically modified crops is now driving up the use of herbicides by about 25 percent a year. Farmers have to use more and more Roundup and add in other herbicides.[48] One need not be a conspiracy theorist to worry about the fact that Monsanto also sells nine other herbicides.[49]

Another synthetic herbicide widely used by farmers and homeowners is 2,4-D, which kills broadleaf plants by causing unsustainable, uncontrolled growth. Canada has approved Dow Chemical's Enlist brand corn and soybeans, which will tolerate 2,4-D. Dow is a major manufacturer of 2,4-D.[50] By 2015, the USDA is expected to approve Enlist. For Americans of our generation, 2,4-D summons memories of protests against Dow for its use, along with its cousin, 2,4,5-T, in Agent Orange in Vietnam. There is serious concern it might disrupt hormones,[51] cause neuromuscular problems, increase the risk of Lou Gehrig's disease, cause male reproductive problems, and make human brains more vulnerable to drug addiction.[52] For days after its application, 2,4-D can vaporize off crops and travel well over a mile in the air.[53] Dow claims its new formula for 2,4-D greatly reduces drift. An uncontrollable problem is that weeds tolerant of 2,4-D will most likely evolve in a few years, just as they did for Roundup.

To summarize: More herbicides are now used than were needed before the introduction of genetically engineered crops.[54] Herbicides may contain toxins that leave residues on or in plants, and those residues go into the cows that eat the crops and hence into you.

Organic farmers use manure instead of manufactured fertilizer, so it might seem that the danger of catching a bacterial disease from cows would be greater when you eat organic food. A pioneering Minnesota study found that although organic food was more susceptible to fecal contamination, it did not appear to pose a "substantially greater risk of pathogen contamination than does conventional produce." Certified organic farms had much less *E. coli* on their preharvest produce than uncertified organic farms. And food from organic farms that aged their manure or compost at least a year (a reasonable period, but longer than is required for federal organic certification) contained about the same amount of *E. coli* and salmonella pathogens as conventional food. Lettuce was the most problematic food tested.[55] As mentioned in chapter 4, *E. coli* can get *inside* lettuce and spinach leaves, a finding that stunned us.

Creative farmers and ranchers often find ways to avoid pesticides. Tom Elliott, former owner of the N Bar bull ranch in Montana, told us he once had a bad problem with leafy spurge, a toxic latex-producing weed that cattle refuse to eat. Dow Chemical had a $50,000 solution: spray his range with a super-potent new herbicide they'd concocted, something that would persist in the environment for five years. So Tom tried it. Spring rolled around and he couldn't wait for all the fresh new grass.

Well, the spray killed brush and trees, but not leafy spurge. And it cut grass production in half. Rather than applying more and more of the poison, Tom let his sheep-raising neighbor graze his animals on Tom's land. The men watched and waited for the sheep to die from the spurge. Instead of dying, they got fat. Problem solved and no more chemicals put in the food chain. Another successful solution to leafy spurge that Tom tried was introducing flea beetles of the genus *Aphthona*. The tiny beetles' larva munched all the way down the plant's thirty-foot roots.

Iowa State University ran an experiment on its Marsden Farm where they compared a typical two-year rotation of corn and soybeans (treated with typical applications of herbicides and fertilizers) to more complex, longer rotation systems that included crops like red clover, oats, and alfalfa (treated with applications of cattle manure but fewer chemicals). The grain yields and overall profits in the more diverse systems equaled or exceeded those of the conventional approach, weeds were well controlled, and the amount of toxins that seeped into groundwater was two hundred times lower.[56]

Organic Is Good for Soil and Cows

For poorly understood reasons, organic farms sequester more carbon in the soil, which helps mitigate climate change.[57] In ways already described, factory farming despoils our air, water, and soil. By locating cows' food far from where cows reside, and by accumulating their waste too far from where it can economically be used to reinvigorate soil, massive environmental degradation becomes unavoidable.

From Clover's point of view, the best thing about the USDA Organic label is that she is guaranteed life in a pasture for the full length of the local grazing season, which must last at least 120 days, and at least 30 percent of her food must come from pasture during that time. But organic cows to be used for meat are exempt from the 30 percent rule during the "finishing" period, when most gorge on organic grain. This giant loophole weakens the organic standard: During the last 120 days of their lives, organic cows may be confined in a feedlot and not obtain any of their feed from pasture.[58]

The USDA Organic label also means that an independent, on-site inspector examines a farm to assure that the treatment of cows meets a threshold of humane handling. For example, organic cows must have their waste picked up frequently; get clean, dry bedding; and have a suitable temperature in their shelter. It's not a pampered life at a Four Seasons resort, but it is vastly better than the spotty coverage offered by state animal cruelty laws, which are the only protection afforded conventional cows. We consider this sufficient reason for buying organic beef and dairy. If a product is also certified by an independent group that monitors humane treatment, odds are that the cows have even better lives.

Organic Cows in Winter

Although dairy cows are invariably portrayed in lush fields of green, come winter even organic dairy cows in colder/rainier climates must move inside. Dairy cows may suffer more than beef cattle from cold because of their thinner skin, lower levels of subcutaneous fat, and lighter coats. Dairy cows allowed outside with wet, just-milked teats can get "teatcicles," which may permanently reduce their productivity. On the other hand, beef cattle that have grown a heavy winter coat, and that get enough food and protection from high wind and constant moisture, can be okay in temperatures as low as the high teens.

One December we decided it was necessary to see for ourselves what an organic cow's life is like in winter. Jay Gordon, executive director of the Washington State Dairy Federation, offered to show us around his own farm. The ground was covered with a thin blanket of snow the morning we met up with Jay at the Rusty Tractor in Elma, Washington. The diner was bustling, filled with weathered men wearing gimme caps with farm-related logos. Jay wore the same sort of cap, and had a short white beard and mustache. A smiling waitress set a plate before Jay that contained a mountain of biscuits drowning in gravy.

Jay is a member of Organic Valley, a nationwide co-op of over eighteen hundred farmers. Members cannot have over five hundred cows. "Let's add more farms," Jay said, "not bigger farms." Because of niche and organic dairies, Washington was the only state to increase its number of dairies in 2011 and 2012. The demand for organic milk is growing fast in China, Japan, and South Korea; in the next seven years, the United States needs to double its production if it is to meet demand, Jay told us. The co-op also expects domestic demand for its product to grow to 5.5 to 6 percent of the total market (in the city of Seattle, organic dairy already comprises 11 percent of the market, the highest percentage in the nation). Organic dairies have an advantage in that their market is growing, they can charge more for their product, and by growing their own feed they are less vulnerable to fluctuations in the price of commodity crops.

The heap of biscuits and gravy grew cold as we peppered Jay with questions. Few things are as unappetizing as cold biscuits and gravy. When we left to ride out with him to his farm, Jay had barely touched his food. He was good-natured about it all.

Even organic cows leave the pasture in cold, rainy months. The Gordon farm in Washington in winter. Photo: Gail Boyer Hayes.

Jay's farm has been in his family for a century and a half. He and his wife, Susan, have four daughters, each of whom has, in turn, served as Grays Harbor County's dairy ambassador (even though the youngest daughter is lactose intolerant). He farms his own six hundred acres organically and manages a few hundred more. All the feed for his 130 or so cows is grown on his farm: grass, alfalfa, clover, corn, and camelina (an oilseed crop with very high levels of omega-3 fatty acids). Oil to power his tractor can be squeezed from the latter so the crop provides more than feed. "When people buy organic dairy, they incentivize the whole system, including farmers growing their own fuel," Jay said. He explained that because it was a La Niña year, the lower temperatures meant there was less protein in the grass he grew. This remark brought home to us the truth that dairy farmers today need to be as well educated about their fields as university professors are about theirs.

Three things about the Gordon farm make it stand out: Jay's intense interest in optimizing forage grain production, the farm's four and a half miles of manure pipelines shared with a neighbor and used to irrigate and fertilize their fields, and the conservation easement on his land that provides habitat for migrating trumpeter swans and other birds and wildlife.

The air was hazy and nippy as we plodded over to the open-sided, free-stall barn to meet his cows. Snow was melting, the ground muddy. We passed a couple of bulls kicking up a sawdust storm in their new bedding. Inside, the gals were lined up, dining. They weren't shivering and didn't appear bothered by the cold. Their legs and undersides were about the same degree of dirty as cows we'd seen in the summer. Their expressions were the usual impassive (stoic?) expression of bovines. The bottoms of cow hooves are smooth and slippery, but the shelter was ice-free.

Jay drove us out into his fields. As we approached a distant field, hundreds of trumpeter swans took flight ahead of us. They soared and swirled as if they were one body, and while they flew they made sounds somewhere between a nasal trumpet and a honk. The largest waterfowl on earth, they are endangered.

Although conventional wisdom says trumpeters are only found in undisturbed habitat, Jay proved to us this isn't the case. In western Washington a quarter of the world's remaining trumpeter swans—some seventeen thousand—dine on farmland. Crop remnants provide food. The habitat value of such farmland is greatly underappreciated. The American Farmland Trust, the Trumpeter Swan Society, Capitol Land Trust, and the National Park Foundation helped to establish an easement protecting this particular wonderful patch of land. Jay still grazes cows on the land used by the swans—if the land isn't grazed, he says, invasive species may crowd out plants such as an endangered lupine that feeds an endangered butterfly. Even an endangered frog thrives in the grazed wetlands.

Chapter Nine

COWBOYS VS. ASTRONAUTS

Cows shaped America's economic system. As the European conquest unfurled toward the west, guns, germs, and steel cleared the way, but it was cows and their keepers that held the conquered ground. When land was used up or ruined, cows and their farmers moved further west. "The existence of an area of free land, its continuous recession, and the advance of American settlement westward explain American development," wrote historian Frederick Jackson Turner in his famous 1893 essay on the significance of the frontier.[1]

The notion of perpetual growth and cornucopian resources begging to be put to use has been called "cowboy economics." Various rationales were created to justify it: manifest destiny, cultural superiority, American exceptionalism. The abundant opportunities available on the frontier led to a widespread belief that anyone with common sense, ambition, and a willingness to work hard could make piles of money.

Midway through the twentieth century, a new metaphor—the earth as a spaceship—began to compete with and slowly displace the frontier metaphor.[2] In 1965, Adlai Stevenson observed in a speech to the United Nations, "We travel together, passengers on a little space ship, dependent on its vulnerable reserves of air and soil; all committed for our safety to its security and peace; preserved from annihilation only by the care, the work, and, I will say, the love we give our fragile craft."[3]

In 1966, Professor Kenneth Boulding explicitly juxtaposed the economics of spaceship earth against those of the cowboy economy. His widely reprinted essay (two years later, Boulding was elected president of the American Economic Association) was hugely influential. The celebrated "earthrise" photo taken in 1968 by Apollo 8 astronaut William Anders lent verisimilitude to the new perspective.

The two viewpoints continue to joust for public support. For cornucopian cowboys, unbounded technological progress has replaced the physical frontier as the new source of unlimited abundance.[4] Spaceship earth adherents—we'll call them astronauts—fear that humans are peeling off Earth's skin of life and replacing it with a new untested skin, effectively betting the planet on a blind leap of faith.

This tug-of-war between competing worldviews has played out dramatically on American farms. When farmers ran out of new land, they turned to the technological frontier: hybrid dwarf plants that produce more grain and less stalk; monoculture crops that simplify harvesting; complex chemicals to fertilize crops and to kill weeds, pathogens, and insects; and expensive, high-tech field machines guided by microchips and satellites. These changes increased agricultural productivity, but at a monumental cost to long-term sustainability. The Green Revolution, with its large machinery and heavy reliance on chemicals, has eroded and compacted soil, poisoned helpful microorganisms in dirt, and replaced diverse crops with patented seeds that farmers must buy anew each year. Now a new generation of farmers and agricultural scientists have begun applying ancient organic lessons that have been beta-tested by Mother Nature over millions of years. Instead of conquering nature, astronaut farmers are studying ecology and seeking to optimize, instead of maximize, their long-term productivity.

Milk, meat, and feed prices are as volatile as the weather. For many farmers, be they cowboys or astronauts, profit margins and savings accounts are thin. Farmers need help to deal with the curveballs thrown their way. Ideally, that assistance, in addition to insuring against the vagaries of weather, would protect them against market manipulation by commodities traders, reward them for taking actions that benefit society and future generations, and penalize them for costs they impose on society at large. Unfortunately, the current program—which pressures farmers to produce too much corn—does the opposite.[5]

Uncle Sam Is a Cornucopian

On a cross-country trip by car in February 2008, during the Obama vs. McCain presidential campaign, we drove from Seattle, Washington, to Washington, D.C. Virtually all the placards we saw when driving through farmlands supported . . . Ron Paul. We were baffled. Why would so many farmers support a candidate who would *eliminate* federal farm support programs and subsidies? Before this trip, to the extent we'd thought about it at all, we'd vaguely assumed that all farmers wanted as much government support as they could get.

So what was going on? To understand this rural anti-government outpouring, we had to read up on the farm bill. Every five years or so, Congress passes a law nicknamed the "farm bill." Farm legislation is voluminous, nearly incomprehensible, dreadfully dull, and designed by high-priced lobbyists to hide things from the public.

Like the rest of society, farmers have their own 1 percent problem. From 1995 to 2010, only 10 percent of all subsidy money went to the bottom 80 percent of recipients; half of farmers received nothing at all. In 1996 the federal government started making direct yearly payments to corn farmers, regardless of their economic circumstances or their farming practices. Farmers no longer had to conserve soil by leaving some land fallow to qualify for payments. So marginal land (with thin soil or steep slopes that quickly erode) was plowed. In 2002, Congress passed a remarkable handout for the biggest megafarms, a deal even the *Wall Street Journal* called "a 10-year, $173.5 billion bucket of slop."[6] By 2012, an additional 15 million acres were planted with corn.[7]

Not until February 2014 did the government finally pass a years-overdue new farm bill, the Agriculture Act of 2014. This new farm bill has a price tag of more than $956 billion over ten years.[8] It eliminates the derided direct payments, substituting additional subsidized crop insurance for (corn and soy) producers, insurance heavily subsidized by taxpayers. The *St. Louis Post-Dispatch* summed it up in an editorial also run in the Iowa's *Mason City Globe Gazette*: "Big corporate farms win. Small farms lose."[9] The *Des Moines Register* opined: "Commodity groups did not fight elimination of these [direct] payments because they would get an even more generous crop-insurance program."[10]

Federal policy is still based on the assumption that there can never be too much corn. The new law again rewards farmers for covering most of America's best farmland with monoculture crops of corn, almost none of which is eaten by humans.

Every farm bill has an enormous impact on the environment. The National Environmental Policy Act of 1969 requires agencies to prepare environmental impact statements (EISs) on proposed "legislation and all other Federal actions significantly affecting the quality of the human environment."[11] It is a tribute to the brute power of the cow-corn axis that no farm bill has ever been subjected to an EIS. Bits and pieces of farm bills have been reviewed, but the nation deserves a comprehensive environmental impact statement before another farm bill is passed. Given the ravenous hunger of our political leaders for campaign contributions from Big Ag, no such EIS will be prepared unless and until a lawsuit wins a judgment requiring it. The National Environmental Policy Act is clear, and the precedents are longstanding. It is time for a full accounting of the impact of the bovine-industrial complex on the American environment.

Although corn growers have seen record profits lately, researchers at the nonpartisan Worldwatch Institute took a close look at exactly who in Big Ag benefits the most from taxpayer subsidies. They found it isn't farmers, but the companies that sell to, and buy from, farmers. The sellers are suppliers of agrochemicals (companies like Bayer, Syngenta, and BASF); farm equipment (dominated by Deere); fertilizer (Mosaic is the biggest); pesticides (Bayer, BASF, Dow); and patented seeds.[12]

Cargill, Archer Daniels Midland, and Zen Noh control 80 percent of corn exports. Four other corporations control 60 percent of terminal grain facilities. Four firms, JBS, Tyson Foods, Cargill, and National Beef Packing, dominate meatpacking.[13]

When you look past the rhetoric and the slogans, it becomes clearer why so many small farmers supported the long-shot presidential candidate who promised to kick the government off the farm. Both Republicans and Democrats have long favored the giant interests that fund their campaigns.

Iowa has made it clear that, come hell or high water, its presidential caucuses will remain the first in the nation. The early caucuses make the state inordinately important to candidates in ways that aren't good for the country. Iowa is not representative of America's economy, politics, culture, or diversity. Iowa is a corn state.

And every four years, Iowa forces candidates for the highest office to make nonsensical, very public pledges to continue to prop up Big Corn. The scheme works ridiculously well.

Between 1996 and 2001, Iowa farmers collected $8.7 billion in subsidies, even though many of them had violated their agreements to conserve soil and water. When fined by the federal government, farmers were able to appeal to county committees, and those committees restored 97 percent of the money. The committees said that the farmers (their neighbors) had been making good-faith efforts.[14] Penalties with a 3 percent enforcement rate are a joke.

One consequence of pouring so much money into subsidizing corn is that the price of high-fructose corn syrup dropped 24 percent between 1985 and 2010. During that same period, the price of fresh fruit and vegetables rose 39 percent. What the USDA calls "specialty crops," such as vegetables, fruits, berries, and nuts, often aren't eligible for federal crop insurance, even though the USDA's MyPlate says such foods should provide half of all calories.[15] The average American adult consumes fruit only about 1.1 times per day, and veggies only 1.6 times per day. Adolescents consume even less fruits and veggies. Residents of California, Oregon, and New Hampshire consume the most.[16]

Even if a specialty crop is covered, insurance isn't available in all counties. To be fair, it's much harder to design insurance (to figure the odds of crop failure) for specialty crops, which are often grown on smaller farms and sold in smaller markets. A single farm may grow many different crops. The complexity of USDA insurance policies discourages many specialty-crop farmers from even applying. For example, barely half of apple acres are insured. There is a trickle of hope that some of our best farmland might switch from corn back to other fruits and vegetables. *New York Times* ace agricultural reporter Michael Moss found some Corn Belt farmers who, for economic reasons, are switching from corn to crops like apples. People in the Midwest have begun to demand fresh local produce (they've lagged the coastal population in this trend). In addition, the price of corn has fallen recently so that (even with subsidies) an acre of corn is expected to earn about $284 in 2014. An acre of apples might make over $2,000, and fancy vegetable operations might even squeeze out $100,000.[17]

Unsurprisingly, Americans who are struggling to make ends meet gravitate toward convenient cheap beef, cheese, and supersized sodas, and they keep getting fatter, less healthy . . . and possibly dumber! A recent study at UCLA found that a diet high in fructose interferes with rats' ability to learn and retain information.[18] The researchers suspect that their findings are also applicable to the brains of other mammals, including humans. (The National Institute of Neurological Disorders and Stroke funded this study; it is not a group with a financial interest in the fructose vs. sucrose feud.) This echoes the finding discussed in chapter 6 that consuming the saturated fat in beef and dairy might also dim your mind. So whether corn is turned into fructose or cow, it's not a health food.

The farmers supporting Ron Paul may have been on to something. Wouldn't it make more sense to focus subsidies on foods that would make us healthier? Our complex, expensive subsidies are not promoting public health, sustainable agriculture, small farmers, or local production.

Maybe it's time for radical change. New Zealand offers an intriguing precedent. In 1984, New Zealand abolished farm subsidies entirely, despite dire warnings about the consequences. Agriculture there did stagger for a few years. But then it surged back. Only 1 percent of farmers threw in the towel—vastly fewer than abandoned their farms in deeply subsidized America. New Zealand's farm output has now increased four times over. Family farmers are reportedly doing well, and so are New Zealand's grass-fed cows. Efficiency has improved. Fertilizer use has declined. New Zealand farmers are competitive on a global scale. Of course, the United States and New Zealand differ in many ways, but our farmers are also alike in many ways. There may be an important lesson in the Kiwi experience.

We Americans spend a smaller fraction of our income on food than the citizens of any other country.[19] But what we spend at the grocery store doesn't reflect our food's true cost. In addition to the hundreds of billions of dollars in governmental subsidies that produced today's agribusiness, the prices in stores and restaurants don't reflect the costs of the harm that high-intensity agriculture "externalizes"—the very real costs to people, soil, rivers, and ecosystems that don't appear on anyone's profit-and-loss statement. But someone has to pay that bill. When land was abundant and fewer chemicals were used, pioneers could toss their wastes into manure piles, gullies, or rivers and nature made most of it disappear. Today, such waste lands in a neighbor's or taxpayer's lap.

Doug Gurian-Sherman, senior scientist at the Union of Concerned Scientists, argues that the economic benefits attributed to the massive scale of confined animal feeding operations are mostly illusory. Because of CAFOs' Himalayan accumulations of cow feces, the limited land around confined-animal feeding operations can't absorb and use all the nutrients in the waste. So the waste runs off or evaporates, polluting the water and air, poisoning people, killing fish, and driving down the values of nearby properties. CAFOs destroy good agricultural jobs, raise health care costs, and eviscerate agricultural communities. This destruction has a real dollar value. Gurian-Sherman calculated some of the hidden but real annual costs of CAFOs.[20]

• Reduction in property values	$26 billion
• Cleaning up soil damaged by leakage from manure-storage facilities	$4.1 billion
• Cost to distribute and apply manure to fields	$1.6 billion
• Public health costs from overuse of antibiotics	$1.5 to 3 billion

Federal subsidies to corn and soy farmers allow them to profitably sell feed to factory farms at prices lower than their true production cost. When smaller herds of dairy cows are pastured on fields large enough to absorb their waste, however, the farmer gets no subsidy. State and federal Environmental Quality Incentives Program projects, which focus on manure disposal problems, also favor CAFOs over pasturing. CAFOs have big manure problems, which small producers don't have, so the subsides flow to the CAFOs.[21]

Welfare Cattlemen

The beef industry has seen less consolidation at the farm level than the dairy industry. As of 2008, almost half of the country's beef cows were on farms and ranches with a hundred cows or fewer.[22] But times aren't good. Cows on ranches of all sizes suffered during the droughts of 2012, 2013, and 2014. The grass dried up, so ranchers had to buy corn or other feed. Many had little choice but to slaughter cows from herds they had spent years developing, cows that hadn't reached their prime weight.

When ranchers have to cull their herds, 84 percent of the cows are sent to one of the four big meatpackers. Capitalism is a contact sport, and these firms drive very hard deals when ranchers are in distress. In the last three decades, 42 percent of ranchers have been forced to quit.[23] A serious effort by the Obama administration to challenge the meatpacking industry with antitrust violations ran into a brick wall in the House of Representatives.[24]

The federal government owns a lot of unsettled land not suitable for growing crops because it's too dry or too steep.[25] Cow and sheep ranchers lease about three hundred million acres of public land in eleven western states. That's roughly the size of three Californias. The leased acreage includes state as well as federal lands, some of it in parks, wildlife refuges, and nature preserves.

Much of this leased land is overgrazed, with consequent erosion, destruction of habitats along rivers, loss of wild species of plants and animals, and the pollution of rivers and streams. These ranchers, some of them quite rich, pay as little as a tenth of what they would pay to graze their cows on comparable private land—a formidable subsidy. A few, such as the infamous Cliven Bundy in Nevada, think it's their right to use federal land without paying even minimal fees. He and his armed supporters chased away Bureau of Land Management rangers who were trying to move five hundred Bundy cows off public land. Bundy owes the government over a million bucks. To avoid bloodshed, the government is pursuing its case against him in court.

Ranchers using federal lands also frequently benefit from low-interest farm loans as well as help with emergency feed, fencing, water tanks, and drought relief. To protect ranchers' herds, every year the government kills hundreds of thousands of predators—wolves, mountain lions, and coyotes.[26]

This misguided land use system is kept alive in large part by a gimmick called an "escrow waiver." Ranchers typically need to borrow a lot of money each year to rebuild their herds. They would prefer not to use unrelated property as collateral for their loans. So the Forest Service allows ranchers to borrow on the worth of their "grazing rights," the renewable ten-year permits they hold on federal lands. The law, however, is clear that grazing is a privilege, not a right; it can be revoked and is not assignable. So how can these revocable permits be used as collateral?

To overcome this legal hurdle, the Forest Service acts as an escrow agent and contracts with a bank to create an escrow waiver. This allows the bank, not the rancher, to "hold" the permit until a rancher repays his or her loan. Meanwhile, the rancher's cows munch away on public vegetation.

Here's the glitch: Banks base the amount of a loan on the value of a permit, which is in turn based largely on the number of cows that are allowed on a given parcel of land. This dubious arrangement routinely results in land being overstocked with cows, because bankers and ranchers put pressure on the Forest Service to keep stocking rates high. Incompetent ranchers are kept in business, overgrazing is endemic, and public land is destroyed.[27]

How important is all this for the American economy? Grazing on public lands accounts for less than 0.1 percent of income and employment in the West, and it provides only about 4 percent of the food that cattle eat.[28] Grazing on public lands may allow us to pay a penny or two less for a pound of red meat at the grocery store. But the saved money slips out of our pockets through environmental degradation, eroding the legacy we leave our grandchildren. The highest and best use of many of these arid, steep, rugged properties is as natural ecosystems.

In the mid-1970s, when the United States had 140 million cows—the most cows ever—*private* lands supported the thundering herd.[29] If the goal of this book is achieved—if the current 93 million cows are gradually reduced to half that number, and raised on grass—it clearly wouldn't be necessary to graze cows on unsuitable public lands.

In 1987, Rutgers University professor Frank Popper and Deborah Epstein Popper, then a graduate student in geography at Rutgers, published a colossally bold proposal to return much of the Great Plains to a "Buffalo Commons."[30] As originally conceived, the Buffalo Commons would have involved returning about ninety million acres of the plains back to the conditions that had prevailed before the westward expansion of the frontier. Involving significant portions of ten states, the proposal received a good deal of exposure, collected some editorial support, and led to the creation of a grassroots organization (the Great Plains Restoration Council) to promote the vision. However, the primary impact of the Buffalo Commons was to inspire people who care about vibrant, diverse ecosystems to begin to think on a much larger scale.

The closest thing to a near-tern realization of the vision may be the creation of the American Prairie Reserve, funded by rich individuals and celebrated scientists who care about preserving grassland for pureblood bison[31] and other native creatures and plants. The group is buying out ranches in northeastern Montana and owns 58,000 acres outright and leases 215,000 acres from the federal government. It hopes to ultimately purchase 500,000 acres and lease or otherwise protect an additional 2.7 million acres. Its scientists and those of affiliated organizations have determined that 3.2 million acres would be adequate to preserve a fully functioning ecosystem, an "American Serengeti" large enough to survive fire, disease, and winter ice events.[32] With large sustainable amounts of prairie vegetation, as well as sustainable populations of bison, pronghorn antelope, burrowing owls, elk, and wolves, the reserve would offer a glimpse of the land as Lewis and Clark traversed it.

Some cattle ranchers in the area oppose the goals of the American Prairie Reserve because they see a threat to their way of life. They worry about bison spreading diseases to their cattle. They worry that the ridiculously low grazing fees that are keeping some of them afloat might rise if the government develops a preference for free-ranging bison over cattle.[33] They might be right. But bison should be preferred. Cattle ranching has seldom been very successful in that dry country, but plains bison thrived there for five to ten thousand years. As climate disruption creates conditions even more hostile to cattle, the American Prairie Reserve may be the best hope to save this unique ecosystem.

Family-run ranches sometimes survive by paying as much attention to the ecology of their land as to the bloodlines of their cows. They can produce enough beef to feed America, and they can do so without treating animals like machines or ruining the land by soaking it with chemicals and causing erosion. Tom Elliott, whose N Bar ranch was large-scale and sustainably run, but profitable, told us: "Factory farms are not something to be proud of as humans." He believes in the inherent superiority of multigenerational family farms because "the land and animals reveal themselves very slowly."

Gary Price owns a small ranch outside Blooming Grove, the 833-inhabitant "Friendliest town in Texas." Unlike his neighbors, he made money on his herd in 2013 despite record-setting drought. As he told *New York Times* reporter Stephanie Strom, he paid close attention to his land and what grew well on it. On small patches of virgin prairie he saw how resilient native bunch grasses were, so he began planting them widely. He saw the sense in sustainable grazing practices. "You're now marketing the grass through the cow," he told Strom. Lack of forage forced neighboring ranchers to cull their herds, but—during three years of intense drought—Price only needed to buy hay for a couple of weeks. Moreover, he supplemented his income with fees paid by hunters to shoot quail nesting in the bunch grass.

Bunch grasses filled the prairies back when bison ruled and are adapted by evolution to survive intense seasonal fluctuations. They have long taproots that find deep water; they are perennials that require no reseeding or cultivation; they grow tall and thick, slow runoff, and provide better water penetration into the soil. Local rains in Price's region have averaged twelve to fourteen inches in recent years, instead of forty-five to fifty inches in the not so distant past. Gary Price isn't counting on the weather ever returning to the old normal.[34]

Smaller Dairies for a Small Planet

Dairies are incentivized to produce more milk than Americans need. Cows' milk production can't be stopped just because the market for milk crashes or production costs soar. Nor can cows stop eating. When demand lags, dairy farmers are reluctant to cull their carefully selected herds.

Fortunately, water can be evaporated out of milk and the dry residue stored or exported. Nonfat powdered milk lasts much longer than powdered whole milk, so that's the only kind bought and stored by the federal government. Dry milk piles up and is stored in giant man-made caves near Kansas City, Missouri, and in cooled warehouses across the country. Although dry milk can be exported, this makes no sense when it is sold for less than its true cost. That's like exporting chunks of our natural resources for free along with the powdered milk.

The goal of government purchases is to ensure an adequate supply of milk and to guarantee dairy farmers a minimum price. However, the Office of Management and Budget found that the program had not been updated as the industry evolved, and that it was not fulfilling its goal. Federal purchases and tariffs on imports keep domestic milk prices higher than world prices, and the USDA isn't required to minimize costs (much less make a profit) when disposing of government-owned dairy commodities.[35] For example, in 2003, the government spent about a billion dollars to buy powdered milk, in addition to storage and transportation costs.[36] That same year, when droughts were drying up pastures and crops in some Great Plains states, the USDA offered to give away 390 million pounds of stored powdered milk to livestock owners to feed to their cows. It sold the powdered milk to the states for just $1 a truckload, and left it to the states to distribute the dry milk. Some states ordered extra and auctioned it off to private brokers. Some ranchers, instead of feeding the milk to their cows, sold it to dealers who marked it up and resold it. Ultimately, the milk that taxpayers had offered to farmers through states for $1 per truckload was selling for hundreds of dollars a ton. Feed dealers and mills had a financial windfall; hungry cows lost out.[37]

Dairy farms are falling in number and growing in size. In California and elsewhere in the West, dairy cows increasingly have been moved into factory farms. Small dairies still dominate in the Northeast, eastern Corn Belt, and upper Midwest, where cows more often

Figure 3. Dairy Farms and Productivity
(decline in farm numbers is offset by rising productivity)

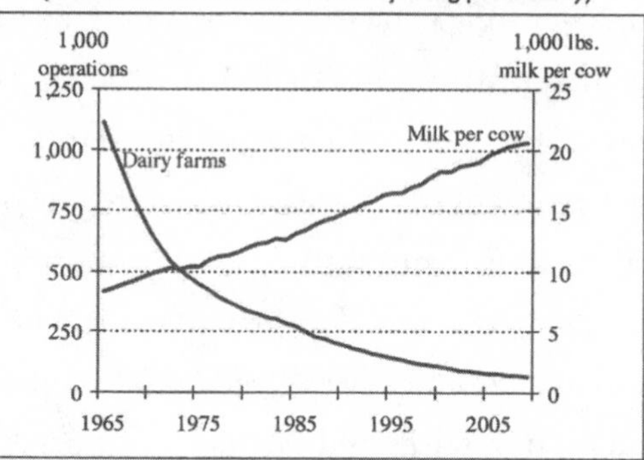

Source: U.S. Department of Agriculture, National Agricultural Statistics Service "Quick Stats."

Notes: Number of dairy operations as of December 31. A milk cow operation is any place having one or more head of milk cows on hand.

graze on grass. But nationally, the number of dairy farms shrank from 1,100,000 in 1965 to 200,000 in 1987 to 65,000 at the end of 2009—a 94 percent decline.

High per-cow production of milk rapidly burns out dairy cows, which are usually slaughtered at factory farms before they are five years old—a quarter of their natural life span. So much stress is put on their bodies that the cows go lame, lose bone, and are prone to mastitis and other diseases.

It's not just dairy farms and udders that are swelling in size. All components of the dairy industry are undergoing consolidation, including cooperatives, processors, and retailers.[38] For example, by 2006 DairyAmerica Inc. (a marketing company comprised of four large producer-owned dairy cooperatives) controlled more than 80 percent of powdered milk.[39] Fluid milk has also consolidated. Because fluid milk is so perishable, and because milk production can't be suspended, dairy farmers are often at the mercy of those who buy their product. They can protect themselves somewhat by joining a cooperative. Cooperatives are partially exempt from antitrust laws. In the early 1900s, only 37 percent of dairy cows were under contract to a co-op. In 2008, 80 percent of milk marketed in the United States flowed through cooperatives, with the top four co-ops handling 40 percent of all milk. In the United States, Dean Foods is the largest processor and distributor of fluid milk. It alone processes a third of all fluid milk.[40]

Federal and state marketing orders determine the price of most milk sold, and that price is set in part by wholesale commodity prices. Consolidation can lead to the manipulation of commodity prices by a handful of purchasers.

Consolidation can also sometimes lead to lower prices for the dairy farmer. A few years ago, giant Dean Foods cut a deal with Dairy Farmers of America (DFA, the largest milk-marketing cooperative) that resulted in a significant decline in individual farmers' share of the profits, yet paid a few men at the top of Dean Foods and DFA many millions of dollars. Shouldn't cooperatives and processors be on opposite sides of the bargaining table? In 2007, two groups of dairy farmers sued. They claimed that the deal was a scheme to reduce competition. Dean Foods settled the cases without admitting wrongdoing: The southeastern farmers got $140 million; the northeastern farmers got $30 million. But many dairy farms had already folded.[41]

Co-ops can be beneficial to dairy farmers if they are well designed. Perhaps the finest example is Organic Valley, the nation's largest independent cooperative of organic family farmers as well as a major distributor of organic dairy. Profits are mostly returned to member farmers. Farmers have a say in setting the price for their milk, and the co-op helps protect them from big fluctuations in price.

There are some economic advantages to size. Largely because of heavily subsidized grain prices, the USDA figures it costs a third less to produce milk at a dairy with a thousand or more cows than at a farm with fewer than two hundred.[42] Fancy equipment, such as merry-go-round milking parlors, automated poop-removal systems, heated and cooled barns (which can increase milk production), and computerized cow-tracking and record-keeping systems require large capital investments that may make better sense at large volumes. Other factors also drive consolidation. Milk is perishable, but it now can be shipped greater distances due to pasteurization and efficient refrigeration. And Americans are eating more cheese and other dairy products that store well and travel more easily than milk.

Small, well-managed dairy farms can be quite profitable, and artisanal dairies can produce high returns on a micro-scale. Prices for milk tend to be higher in regions with smaller dairies, but that is because the smaller dairies internalize more of their costs instead of dumping them on society, and because they don't benefit as much from huge federal subsidies.

Organic dairies benefit from the premium paid for their milk. Organic farming is one of the fastest-growing segments of agriculture and by 2012 comprised over 4 percent of food sales. Sixteen percent of milk sales were from organic dairy farms.[43] The trend lines are encouraging, though we have a long way to go.

A National Research Council study concluded, "Alternative farming practices typically require more information, trained labor, time, and management skills per unit of production than conventional farming." At the same time, alternative farmers spend less on large machinery, fertilizer, pesticides, and antibiotics.[44]

Many small farmers, especially those within driving distance of cities, are getting creative in finding new sources of revenue. We experienced this firsthand when we enjoyed a fall trip with our granddaughter to a farm in Virginia that was an easy drive from Washington, D.C. For the Halloween season, the farm sold pumpkins and had a corn maze and a hay-bale slide. The place was packed with city kids meeting farm animals and learning where their food came from. Like an increasing number of farms, it also sold food on-site.

Although the game is currently rigged against them, small-farm families can compete successfully by keeping up with new knowledge about best farming practices and applying their learning to the particulars of their land: its soils, its warm spots, its wet places. They can do so by conserving and improving their soil by rotating crops and cows, planting cover crops, and using their cows' manure for fertilizer. In China, some plots of land have been farmed productively for thousands of years without losing their fertility. Smaller farms can develop networks though farmers' markets and social media to market their products directly. Farmers' markets increased in number from 1,755 in 1994 to 4,685 in 2008.[45] Beginners can find helpful information on the Internet, from targeted publications like *Modern Farmer*, and from community-supported agriculture organizations.[46]

A Spaceship Earth Cowboy

Bryan Ulring is the manager of the J Bar L Ranch, a twenty-thousand-acre spread near Dillon in southwestern Montana. He's also the owner ("chief ruminator") of Yellowstone Grassfed Beef. Denis accompanied Bryan on visits to Seattle restaurants where this shy man pitched chefs and staff on his organic, grass-finished beef. Even though the audience was composed of foodies, Byran kept the lens wide and pulled them along. It didn't hurt that Bryan is as handsome as a movie-star cowboy.

After covering the *terroir* and health benefits of his beef (Bryan is college-educated), he described how he manages for biodiversity. If wolves or bears show up, he just moves cattle to another part of the ranch and patrols more instead of shooting the predators. He told of keeping cattle out of key habitat areas during nesting seasons for sage grouse and trumpeter swans. And he described in detail how he confines his cattle in tight clusters where they graze intensely for a few days before he moves them along. He and his ranch hands use portable, solar-powered, single-strand electric fences instead of six-strand barbed wire. The fences keep the cattle in but allow wildlife to pass over or under. "This intensive, holistic approach mimic the buffalo that used to graze on what is now the J Bar L," Byran said, "herding them close together for protection and moving on when predators get too bold."

Leah Hair, a longtime friend who has visited the J Bar L repeatedly to ride with the ranch hands, swears that the grass is thicker every time she returns. Byran confirms that. "In some places, we have three or four times as much grass as we had before." The operation is more profitable and the supporting ecosystem far richer than when Byran began the holistic approach. "We manage for biodiversity, and it works for the cows, too," Byran says.

Like other successful, innovative ranchers, Bryan pursues every angle that might contribute to success. He has superior beef that has given weight to his brand name, and he also has the drive and entrepreneurship to put on his Stetson and go wherever he must to find a market. He runs a small guest operation ("Don't call it a dude ranch!") on the side. "Welcome to the Middle of Nowhere. . . . J Bar L guests disconnect with the world and reconnect with what's really important: Nature. Family. Friends. Food," says the ranch's website. Guests eat well and sleep comfortably but can pull their full freight if they want: Castrating, branding, removing ovaries, and ear tagging can all be in a day's work. The dozen or so guests at any time make a real contribution to the ranch's financial viability. And share an authentic experience they never forget.

A Good Way to Live and Raise a Family

Organic Valley's website has a feature that allows you to enter an area code and the site takes you to the page of the closest member farm. We were booked to attend a fund-raising event at a winery in rural Oregon, so we decided to make a weekend of it. We entered in the winery's zip code, and up popped a group of farms. Jon and Juli Bansen of Double J Jerseys looked friendly. We cold-called them and asked if we could pay a visit. It was that random. Jon said to come on down.

We packed our muck boots and headed south, across the mighty Columbia River, to Yamhill, Oregon. It was early fall, perfect cow weather. Not too wet, not too dry. Not too hot, not too cold. Stratocumulus clouds wafted across a blue sky.

Penny, the farm dog, ran out of the unpretentious ranch house to sniff out our intentions. Jon wasn't far behind. A good-looking guy in beltless jeans, a red t-shirt, and glasses, left hand thrust deep into his pocket, he had the sort of unforced smile that generates instant credibility.

No cows were to be seen. Swallow houses were everywhere, however, one kind for cliff swallows, one kind for barn swallows. Where you find cows, you usually find flies, and swallows dine on flies. "Cleanliness is job one," Jon says. "Leave a mess anywhere, and the flies will find it." The cliff swallows get pastel wooden houses strung along a wire, the barn swallows have rounded mud-colored houses packed under the rafters of the clean, but empty, barn. Thousands of swallows call these birdhouses home, but only for fly season. Like migrant farmworkers, they follow the seasons.

Denis asked, "Didn't it cost a lot to put up all those birdhouses? How does that generate revenue?"

"If getting cancer from sprays isn't an economic issue, I don't know what is," Jon replied. The birdhouses save money that would otherwise be spent on pesticides, and perhaps eventually on medicine. Jon, an inveterate tinkerer, is also looking into a walk-through vacuum system to suck flies off the cows.

In college, Jon majored in biology. He knew the habits and properties of every bird and plant we passed. A third-generation dairyman, he had spent six years working on his dad's dairy farm before he and Juli bought their own farm in 1991 and began transitioning it to organic. Like every successful farmer we met, he's smart and adaptive. You can't just wake up one morning and declare yourself an organic farmer. First, the land has to be managed organically for three years. The cows have to be born from an organic mother and continuously organically managed. There's lots of paperwork, more than many farmers are used to. But there are also funds to ease the transition.

It's downright scary to convert a farm into an organic dairy because Big Ag, backed by its minions at the nation's agricultural colleges, has thoroughly conditioned farmers to use chemicals. "If you don't use all their lotions and potions you fear you'll fall off a cliff," Jon says. The Double J has been organic for over a decade now, and it has worked out well financially. To buy the farm, Jon and Juli went a half-million dollars in debt. But they worked hard, saved, lived frugally, paid off their debt, and now have an upper-middle-class income. "Chasing *stuff* doesn't do much for me personally," Jon says. "We buy good stuff we really feel strong about and don't buy the rest." They live comfortably, but beneath their income.

Even on an organic farm the cows can't stay outdoors year-round. The cows don't like heavy rain. Or cold. And pastures would be stomped to ruins when the dirt is mushy. By March, however, the gals usually get back outside, even though Jon has to sleep with his window open so if a downpour starts he will hear it and can get the cows back into the barn. The work is unending and hard. But there's a pleasing rhythm to it. He calls the in-and-out-of-the-barn during early spring "the dance." Then there's the cows' daily pattern of getting milked, grazing, lying down and chewing their cud, grazing, lying down, grazing, and getting milked again.

The Double J employs a man to help out, but Jon and Juli do most of the work themselves. Jon even inseminates the cows himself. "I think they prefer me to a two-thousand-pound bull jumping on them." He's getting sperm from New Zealand, because that country has been breeding for grass utilization by cows. "The world forgot what a cow is supposed to eat." Most of his Jerseys are also part Friesian, Ayrshire, or KiwiCross.

The Double J milks about two hundred cows, Jon told us. But we still hadn't seen a single adult cow. Or even a cow pie. It gradually became apparent to us that Jon had a quiet flair for showmanship, showing us the mundane aspects of his farm first and saving the cows—and a special cow named Rosie—for last.

The Double J uses intensive, rotational grazing, shifting the cows among three-acre paddocks. This is made possible by movable electric fences. The system was arrived at after years of keeping detailed records. Every part of the farm can be reached with a traveler irrigator; liquid manure is put into a lagoon and then gunned onto the Bansens' 175 acres (they lease another 300 acres from neighbors). The system saves them $50,000 a year in fertilizer costs.

Juli's specialty is caring for the calves. Jon finally showed us some cows—two month-old calves. It takes a hard heart not to go, "Awww," when looking into those trusting, big black eyes. Calves stay on a milk diet for about a month, waiting for their monogut to develop the four chambers it needs to digest grass.

We walked down gently sloping land toward the North Yamhill

Most organic farmers grow all or much of the food their cows eat. Jon Bansen in the Double J's field of Sudan grass. Photo: Gail Boyer Hayes.

River. Big-leaf maple, ash, and white oak dotted the pastures. The Double J irrigates with river water. We passed a field of Sudan grass, another of plantain. Sometimes the Bansens plant chicory or orchard grass. They aim for minimum tillage. When Jon imports feed, it's organic. He said he gives his cows only a little grain. A clover rotation is essential, he said, to keep up the nitrogen content of the soil.

We learned that cows aren't picky eaters: mallow, dock, and raspberry shoots all become cud. They prefer perennial rye grass and love clover, but too much clover (unless it has been dried) gives them bloat. On the other hand, cows won't touch dog fennel (chamomile). If you don't find this sort of information fascinating, organic dairy farming probably isn't your calling.

As we headed down-slope, we spotted a lone cow off by herself in a pasture. "That's Papaya," Jon said. "Looks like she's dropped." His cows sometimes like to give birth off by themselves, in a quiet place. At Papaya's feet lay a folded-up calf. Buzzards spiraled in the sky above, hoping for a bite of the nutrient-rich placenta before Papaya ate it all. When the calf didn't immediately get up, we waited

Protected riverbanks at Double J Jerseys. Some commodity crop farmers don't like to leave any ground unplanted. Photo: Gail Boyer Hayes.

with ranch dog Penny while Jon walked over to the new mother and helped the calf to its feet. We had the pleasure of watching this new creature take his first look around the world.

We left the new family and continued downhill. Gail finally asked, "Where are the cows?"

With the laconic skill and timing of a gifted showman, Jon pointed to a small bridge ahead. "Across the river. They should meet up with us before they get to the bridge."

How could two hundred cows be so quiet? The Yamhill River here isn't very large; we crossed it on a serviceable bridge about fifty feet long. Midway across we paused to note with approval the lush vegetation that protects these riverbanks from cows and erosion.

Then cows seemed to appear from nowhere. Two hundred of them, neatly lined up one or two across, walking silently toward us. They were beautiful and healthy-looking. We were struck again by how a cow's hide drapes over its frame like thinly padded upholstery, with shoulder and hip bones poking up like tent poles. The cows were curious about us; Jon says Jerseys are the most curious of cows. A dozen or so left the line and walked over to us, warily turning their heads a bit to one side to get a better look. Jon warned us not to look them in the eye too long.

We met Tory, Taffy, and Tufu, Pinto, Salsa, and Flurry. They were fifty shades of tan, with an occasional black. Their barrels (rib cages) were wide, which let them take in a lot of grass and produce a lot of milk. The Double J's goal is forty pounds a day of 5 percent butterfat milk from each cow. When breeding cows, Jon looks for genetics that complement the type of grass he grows.

We followed the cows back to the milking barn. Each cow backed into a stall and helped herself to a calming nibble of oats. A farmhand wiped her teats with iodine and hooked her to a milking machine.

Jon comments that organic farming is not a lifestyle for the lazy.

Double J Jerseys' ladies head for the milking barn.
Photo: Gail Boyer Hayes.

Record keeping takes a lot of time. In the winter it rains, and dirt turns into mud. Cows move indoors and their poop has to be scraped out and put into the lagoon. Food has to be brought to them. Cows need to be milked even when it's cold and rainy. New calves need coddling.

But for the right sort of person, organic dairy farming is a terrific job. "There are tons of opportunities out there," Jon says, "because many dairymen are close to retirement age. But most kids don't see a future in working that hard." His advice for the undeterred: If you don't come from a dairy family, you need to work for a dairy for about five years and pick up the necessary skills. "It's a good life," Jon said. "And a good way to raise kids."

With justified pride, Jon finally introduced us to Rosie, a beauty who has already lived about three times as long as the typical dairy

Award-winning Rosie, the highest lifetime milk producer in Oregon. Photo: Gail Boyer Hayes.

cow. (It isn't unusual for organic cows to live much longer lives than conventional cows. Stonyfield cows, for example, produce for ten or more years, some even up to twenty years.) Rosie had just won an award for being the highest lifetime production Jersey in Oregon. This cow has charisma! Her ears appear to be rimmed with mink. On Rosie, even flies look like black diamonds. A gentle pet, she allowed us to stroke her, feed her hay, and even look her in the eye.

A farmer can't afford to get too sentimental about his or her cows. But Rosie has clearly claimed Jon's heart because she has given his family so much. She has been granted the rarest of tributes to a dairy cow: the right to live out the rest of her natural life on the organic Double J Jerseys farm—with full benefits and no more inseminations.

Chapter Ten

DON'T BE CRUEL

A six-year-old brown dairy cow named Yvonne thrilled cow-loving Germans in 2011 when she escaped minutes before she was to be slaughtered. Apparently more muscular and resourceful than most dairy cows, Yvonne lived free in Bavarian forests for over three months, attracting fans from around the world. A tabloid offered €10,000 for her safe return. Animal rights advocates clashed with hunters who had been given permission to shoot her. The public cried out and the permission to hunt was rescinded. Helicopters, infrared cameras, and a handsome bull were used to try to track and lure Yvonne. All failed.

After ninety-eight days on the run, loneliness for other cows finally led to her capture: She was spotted amid a farmer's herd. Still, it took many men and a tranquilizer shot to control her. Yvonne is living out the rest of her natural life in a sanctuary. Confined, but a celebrity. Why did she so capture people's hearts? Maybe because many of us are uneasy about how we treat cows. Maybe because, penned in our modern cubicles, we see a bit of ourselves in Yvonne.

Humans bred cows to be dim-witted and placid. Given the large size of cows, this was a necessity if we were to coexist in close proximity. But IQ isn't the same thing as the ability to feel fear, pain, happiness, or anxiety. That ability is called sentience. Cows still have a lot of sentience. We met a farmer who denied this, who claimed cows were about as sentient as a Barcalounger. But studies measuring the stress hormones in cows' blood have repeatedly disproven this notion. For example, stress hormone levels double or triple when cows are prodded too much or slip on slick floors. If cows are crowded too closely together, it causes psychological stress for subordinate cows that have trouble finding a place to lie down or feed. Stress leads to illness, just as it does in humans.

There are a great many stories about cows expressing emotion. One is the tale of the dairy cow that walked over seven unfamiliar miles to find the calf that had been sold and taken from her.[1] The hormone oxytocin (boosted by the birthing process) fosters emotional attachment in both humans and bovines. Call it motherly love, or call it instinct, it probably "feels" about the same in any mammal. If a calf is stillborn, a cow will nose it, roll it over on its stomach, lick it,

Curious, alert young cows. Cows feel stress, love, fear, pleasure, pain, and contentment. Photo: Liz Salim.

smell it, prod it. She might wander away but comes back and repeats her actions. Finally her back arches, her head lifts, and she howls.[2]

Cows organize themselves into complex hierarchies within herds, seemingly based on age and strength, and choose a leader and a dominant cow (they might not be the same). A cow prefers not to feed right next to a cow of higher rank. Re-sorting cows into new groups raises plasma levels of the stress hormone cortisol, which may lead to increased aggression.

Cows form friendships and enemies within a herd and will "calf-sit," taking turns watching over a bunch of calves. They spend a lot of time grooming and licking their friends. Cow "pairs" tend to feed at the same place at a feed bunk.[3] If a cow notices that a cow friend is upset, that cow will eat less, seemingly showing empathy. Like Yvonne, cows panic and try to escape when they sense other cows being harmed.

A study done at the University of British Columbia found that a dairy calf put in a pen with another calf (after being weaned from Mom), seems smarter than a calf kept alone in a pen or hutch. When added to the herd, calves that had a companion more quickly figured out how to use the feeder, for example.[4] We'd guess they were also happier, having a chum.

Cows even show delight in learning new things. When a cow learns to press a panel to open a gate to get access to food, her heart rate increases and she moves faster toward the food than control heifers that don't have to press a panel to get the food treat.[5]

The late Bud Williams, a cattle handler from Alberta, Canada, had a rep as a "cow whisperer" extraordinaire. His secret? Work with cows' built-in instincts and stay mellow. No shouting, rapid movement, whip cracking, or unexpected moves. When you want to move a herd, the cows should be a tad anxious but not scared. Don't rush things. To trigger their clumping instinct, act like a predator and begin moving in a wide arc opposite where you want them to go, staying just at the edge of the zone where they might take flight. Keep moving back and forth at a walker's pace. If needed, give a cow a meaningful predatory stare. Ignore outlying stragglers. When the cows are in a bunch, move in a little closer. Keep the pressure light. A leader cow will eventually go through the gate or into the desired fresh pasture, and the others will follow.[6]

Farmers have long noted that anxious cows don't give as much milk or put on weight as well as calm cows. Back in the day, dairies advertised milk from "contented cows." A study at the School of Psychology, Queen's University Belfast, found that stockpersons whose attitudes were more empathetic toward their cows, who liked their jobs, and who had fewer negative beliefs about cows, got more milk from their animals.[7] Similarly, researchers at Newcastle University in England interviewed 516 dairy farmers and learned that when a cow was given a name and treated as an intelligent individual possessing a range of emotions, the cow could increase her annual production of milk by nearly five hundred pints. So a happy cow is a cash cow. Smaller farmers often refer to their cows as "our ladies" and "part of the family."[8] Of course, naming cows probably raises the farm family's stress levels when the time comes to slaughter the cow. People, too, have feelings.

What It's Like to Be a Dairy or Beef Cow Today

Proponents of factory farming may point out that confined indoor cows are safe from wolves, hail, poisonous weeds, cownappers, and black helicopters. True, but those aren't major worries in most dairy cows' lives today.

Clover is our archetypical dairy cow (although on a factory farm she would have only a number). She is a Holstein, one of those big, buxom, black-and-white beauties. After nine months of gestating inside her single working mom, Clover is born on a farm with over a thousand cows. Chances are she will be allowed to stay with her mom only a few hours, and then be fed milk or formula from a bottle or bucket.

Little Clover is put in a separate stall, pen, or hutch for several weeks before joining other calves in a group pen. (A hutch may be one of those white plastic mausoleum-like structures you see lined up near barns). Although being with other cows increases a calf's comfort level, Holsteins are far more likely to catch a disease than are beef cattle, so it is chancy to put calves in with other cows too soon. Crowding (as well as her genetic weakness) will make Clover prone to infections all her life.[9]

Breeding dairy cows for higher milk yield has had some serious and unfortunate side effects: Clover is weaker than a beef calf and has more trouble walking unassisted. She also has a reduced life expectancy, may have fertility and metabolic problems, and is more susceptible to mastitis and other production-related diseases.[10] Livestock behavior expert Temple Grandin notes that breeders have focused so intently on maximizing milk production that they've created a fragile animal that couldn't survive in the wild.[11]

To make Clover grow faster, she is fed too much grain and too little roughage. She may grow so fast that her skeleton and hooves fail to harden well. The high-grain diet also causes bloat and constant stomach problems. "If you think of the cow's metabolism as being like the rpm tachometer in a car, the industry is running dairy cattle in the red zone and burning them up," writes Temple Grandin.[12]

Clover's short life on a factory farm will often be lived entirely indoors. She will never see the sky or green grass. She will be milked two or three times a day. She will usually live in a freestall barn that she can move around in and mix with other cows, but where she also has access to a separate stall in which to lie down. Being able to lie down is extremely important to cows. If she does get to go outdoors, in the dry Southwest it probably means she will be in a "dry lot" of unpaved bare dirt, an invitation to lameness.[13] Concrete floors are also associated with lameness. A few factory cows now get soft rubber flooring—this makes it easier for them to walk around without slipping.

When Clover is around fifteen months old, she'll be artificially inseminated.[14] The sperm will most likely come from one of a relatively few popular bulls (dairy farmers leaf through semen catalogues with the same rapt attention to detail that young geeks reserve for *Macworld*). This extreme selectivity of bulls is leading to a decline of genetic diversity.

Clover will be milked until the last two months of her pregnancy. Two to three months after she gives birth to her first calf, she will again be artificially inseminated. A lot of calcium is needed to make both milk and a calf, and some of it is stolen from Clover's bones. She will have only two to three lactations before she's "spent" and slaughtered. That means she'll live less than five years, a quarter of a cow's normal life span. A Droimeann cow born in Ireland and named Big Bertha holds the Guinness World Record for longevity. She lived to nearly fifty. (She also holds the record for producing the most calves: thirty-nine.) Maybe Big Bertha's longevity can be attributed to the occasional shot of Irish whiskey she was said to enjoy, to calm her nerves before a public appearance.[15] In the United States, regulations aimed at reducing the chance of mad cow disease encourage the slaughter of cows before they reach thirty months of age (chapter 7). Maybe because cows now have such short lives, Guinness has dropped the categories in which Big Bertha excelled.

When the time comes, Clover will be transported to a slaughterhouse while she can still walk. Weakened dairy cows are often injured while being trucked to slaughter. If she becomes a downer cow and can't walk, it will be very difficult, or impossible, to treat her humanely once she leaves the farm. About 17 percent of our beef comes from exhausted dairy cows.[16] They usually end up being made into hamburger.

About half the calves born to dairy cows are, unsurprisingly, male. A few male dairy calves are kept and used for breeding, or castrated and raised for beef as steers. (At one time dairy and beef cows were often interchangeable, but today's specialized dairy cow makes a second-rate beef cow.) Around 15 percent of male calves are killed when two or three days old and sold as "bob veal." Male calves are traditionally put in a "veal crate," tiny confined spaces they can't move around in (so they won't develop tough muscles) and where they experience sensory depravation. Baby veal cows get sick living under these conditions and require even more antibiotics than other cows.[17] Residues of these drugs may remain in their meat. We found it heartbreaking to approach tightly confined, tethered calves and watch them try to twist their heads around so they could just look at us.

If a calf is to be raised for "white" veal, he is fed only formula milk and is killed when he's about five months old. The goal is to produce pale flesh with little connective tissue. If he's fed grain or hay, he may have to endure up to nine months before he's killed for "pink" or "red" veal.

In the United Kingdom and the European Union, the tiny stalls are illegal. The American Veal Association issued a policy statement in 2007 encouraging the phasing out of close confinement by the end of 2017; it instead encouraged the use of group housing.[18] In August 2013, however, the Veal Farm website, funded by the Beef Checkoff, was still ambivalent. On one hand, it promoted group housing for veal calves; on the other, it still extoled the benefits of tiny growing stalls.[19] Five states have banned restrictive veal confinement. Encouragingly, a 2012 survey of members of the American Veal Association found that 70 percent of members said that by the end of 2012 they would be raising their calves in group pens where they could interact with other calves.[20]

In 2009, the Humane Society obtained videotape of veal calves (some with umbilical cords still dangling) being given electric shocks, kicks, and slaps at a Vermont slaughterhouse. A calf with its head half cut off cried out. The horror goes on from there.[21]

Some farmers try hard to raise their veal more humanely. "Free-raised veal" calves get to eat pasture grasses and aren't given antibiotics. A "certified humane" seal offers assurance that the veal calf has not been subjected to gratuitous violence before its death. Usually, however, there's no way to tell how the veal you find at the grocery store was raised.

Fewer and fewer people are choosing to eat veal. In 1944, per capital consumption was 8.6 pounds of veal. By 2009, it had fallen to 0.4 pounds.

To our surprise, we learned that cows (steers) raised for beef have better living conditions, overall, than factory-farmed dairy cows. Ferdinand, our archetypical male beef cow, is born to a Black Angus. (Female cows are also raised for beef.) His mother will eat the placenta after her calf is born to regain its nutrients and so predators are less likely to know she has given birth. (This is the natural exception to the rule that cows are vegetarian.) Then she will suckle Ferdinand. As he weans, he learns to eat grass at his mom's side.

There are various methods of growing ("backgrounding") beef cattle. Some breeds are better suited to a long growing program, some get none or almost none. The latter may be "weaned on the truck" on the way to the feedlot and are more likely to get sick after they arrive.[22] Ferdinand, however, stays for a while on the ranch. At around two months old he's castrated, vaccinated, and branded. After six to ten months, he is taken from his mother (she will bellow her heart out for a while) and grazes in a pasture. When fall comes and there's less fresh grass, he gets supplemental feed high in roughage. After he's put on enough weight, he'll probably be sold at an auction and moved to a feedlot.

During his four to six months in the feedlot, Ferdinand will gorge and pack on the pounds.[23] Some beef cattle are fed nothing but an unhealthy diet of grain to fatten them quickly. "Frankly, it is a race between fattening the animal and killing it," writes M. R. Montgomery, in *A Cow's Life*.[24] No matter the breed, cows can't digest a high-protein, high-carb diet without getting indigestion. So Ferdinand (like Clover) is routinely given sodium bicarbonate or other alkaline/buffering agents along with his chow. He is also given antibiotics and steroids.

Heat stress and mud are constant problems in feedlots. Most feedlots are located on the High Plains, where there's less rain to create mud (but an aquifer underneath to pollute). At best, it's not a pretty picture, and "best" is rare. More than 40 percent of fed cattle come from feedlots containing thirty-two thousand cows or more.[25] When tens of thousands of cows are crammed into the same feedlot, they can't be picky about poop when they look for a place to lie down. Sleeping in feces is as bad as it sounds.

Something called "buller steer syndrome" is far more common in feedlots than on the range. This is where a bunch of "rider" steers mount another steer repeatedly until he's badly hurt or dead. It's hypothesized that the steroid growth hormones given steers—particularly, the sloppy insertion of steroids so the hormone is released all at once—might be partly responsible.

As soon as Ferdinand packs on enough weight, he is slaughtered. Few beef cattle live beyond thirty months.

Elective Surgery on Cows

Humans remove several parts of cows to make the animals more pleasant to work with. All of these practices need reforming to lessen the pain to the animal undergoing remodeling. One practice, tail docking, should be banned in America as it has been in much of Europe.

A bull recognized as having exceptionally fine traits might be kept intact for breeding. But the vast majority of bulls, like eunuchs in a sultan's court, are castrated. A castrated bull is called a steer. A bull is separated from his bull-hood about fifteen million times a year in America. Steers don't get as muscular as bulls, and their fattier meat commands a higher price. The surgery also dramatically reduces Ferdinand's production of testosterone, which makes him less likely to fight other males or smash up fences. Without castration, buller steer syndrome would likely be even more common.

Castration is painful. We know this not just from the bellowing but because researchers find a rise in blood cortisol levels following castration. Using an anesthesia or nonsteroidal anti-inflammatory drug reduces the pain. But a survey by Kansas State University found that only a fifth of veterinarians used an anesthetic or analgesic.[26] Cowboys and farmers, who do a lot of the castrating, almost never use pain-reducing drugs. An anesthetic requires several minutes to take effect and some skill to administer. Some cowboys argue that the stress to the calf from being separated longer from its mother outweighs the benefit of making the surgery pain-free.[27]

Several techniques are used to castrate bulls.[28] Many cattlemen prefer surgical removal, which can be bloody. An advantage in some minds is that this method provides edible calf testes. ("Rocky Mountain oysters" are generally peeled, pounded flat, and served deep-fried.) A bloodless, but very painful, method is to put a clamp or rubber ring around the base of the scrotum, cutting off the blood supply. Eventually the organs are reabsorbed or fall off. Then there's a device called a Burdizzo, a large pincer/clamp that breaks blood vessels leading to the testicles. It is also possible to inject toxic agents, such as lactic acid, into the testicles, or to repeatedly inject hormones that inhibit testosterone production.

As we have seen, a cow is said to be polled when it has had its horns removed or was bred not to have horns. (A poll is the ridge between or just behind a cow's ears.) Both male and female cows grow horns. Humans have been selectively breeding cattle to get rid of horns for a long time. Angus beef cattle, for example, now have no horns. But many other beef cattle and nearly all dairy cows grow them. Horns can injure udders and eyes, maim farmers, and are a major cause of "carcass wastage."

Until about two months of age, little horn "buds" are not yet attached directly to the skull, so the cow is usually "disbudded," at this stage. Disbudding can be done with a butane dehorner, with a hot iron, by scooping them out, or by using chemicals such as "dehorning paste." All methods hurt. Horns can also be removed from adult cattle. This is much harder on both the cow and on the human attempting to do it.

Although some countries require measures to make the procedure less painful, dehorning isn't regulated in the United States. Using a combination of sedatives and a local anesthetic before dehorning, and a nonsteroidal anti-inflammatory medication afterwards, can significantly reduce the pain.[29] Most North American cows are dehorned without any anesthetic. A survey of 113 dairies in the Midwest and New York found that about 67 percent of calves were dehorned by hot iron, 9 percent by gouging, 10 percent by paste, and 3.5 percent by saw. Only a bit over 12 percent of dairy owners used anesthetic.[30] Not convinced that dehorning hurts? People for the Ethical Treatment of Animals (PETA) has a video that will persuade you.[31]

Dairy cows' tails are sometimes amputated (docked) to make them only a foot long, because some dairy farmers think it's more hygienic. People milking cows don't like being whipped by urine- or feces-covered tails. But scientists have not found any food safety problems, or cow cleanliness issues, caused by intact tails. Tail docking is usually done without anesthetic. A tight rubber ring is put on the tail and left there until the end of the tail dies and falls off or is cut off. Cows use their tails to swish away flies and possibly to send signals to other cows, and cows without tails suffer more from fly bites.

Several European countries have banned docking, as have the states of California, Rhode Island, and Ohio. The American Veterinary Medical Association (AVMA) and the Canadian Veterinary Medical Association oppose routine docking. According to the AVMA, temperature sensitivity and the presence of neuromas (tumors made up of nerve cells) suggest that chronic pain may be associated with the procedure.[32] Even the large National Milk Producers Federation opposes this procedure.[33] Nevertheless, a slight majority of dairies are said to shorten their cows' tails.[34] We think this issue is clear-cut: Docking is bad.

At the slaughterhouse, a cow is usually stunned with a "captive bolt pistol" or a gunshot to the head. This should make the cow immediately unconscious. Within about fifteen seconds—before it regains consciousness—the cow is ordinarily bled to death. This is not something most omnivores like to think about. But when slaughtering is done correctly, cow suffering can be minimal.

In the late 1950s, the Humane Slaughter Act was passed. This federal law requires animals to be unconscious when they are killed. The USDA's Food Safety and Inspection Service is supposed to enforce the law. But enforcement is very spotty. The law provides exemptions for slaughtering preferences based on religion. Furthermore, it applies only to slaughterhouses, not to farms. Adding to cows' misery is the long ride sometimes needed to get them to one of the few remaining slaughterhouses.

Slaughtering is often not done correctly, and cows are subjected to revolting mistreatment. When stunning is flubbed, the poor creature may be skinned and disassembled while still alive and conscious.[35] As related earlier, a 2008 undercover video made by the Humane Society showing mistreatment of cows at the Hallmark/Westland slaughterhouse in California was leaked to the media. Hallmark/Westland was then the second largest meat processor in the country. The Humane Society and the Department of Justice brought a civil lawsuit against Hallmark/Westland. The suit was based on fraud—as a supplier of meat for the national school lunch program, Hallmark/Westland had signed a contract promising humane treatment of animals. Late in 2012, Hallmark/Westland reached a $500 million settlement with the plaintiffs. It was an empty victory, however, because the meat company had gone bankrupt.[36]

The Hallmark/Westland situation was not an isolated incident. A videographer documented employees at the E6 Cattle Company in Texas bashing in cows' heads with hammers and pickaxes.[37] Another videotape, this time at an Idaho dairy, showed a downer cow, a chain around her neck, being dragged over a concrete floor by a tractor, and a cow being beaten because her leg was caught in a milking stall.[38] Compassion Over Killing (COK), a vegetarian group, obtained video of half-stunned cows at Central Valley Meat in California being hung by one leg prior to slaughter.[39] Other videos show cows being shot many times in the head because of shooter ineptitude, and being suffocated, stabbed, and lifted by their tails.[40] Such visual documentation has a powerful emotional impact. Few consumers want to eat beef or cheese from farms that treat animals with appalling barbarism. When Mercy for Animals released a video showing cruelty at a Wisconsin dairy farm, DiGiorno pizza and its cheese supplier stopped buying cheese from that farm.[41]

Temple Grandin, whose autism allows her to better see the world from a cow's viewpoint, has spent much of her life trying to improve the treatment of cows. Grandin visited twenty-one commercial slaughter plants and watched the captive-bolt stunning of 1,826 fed steers and heifers and 692 bulls and cows. At seventeen plants, all cattle were insensible when hoisted onto the bleed rail. But at the other four plants, eleven cows regained sensibility en route to "disassembly."[42]

A professor at Colorado State University, Temple Grandin has discovered many ways to reduce cows' fear. A simple thing to do, for example, is to fix cardboard to the barred sides of a squeeze chute so that the cattle don't have to see the people standing close to the chute. She suggests that ranchers pay attention to how much manure cows release into a chute. "I tell ranchers that the cattle poop is less because they are no longer scaring the s*** out of them." But when she returns to a facility where she previously trained employees, she often finds they have slipped back into their old ways: prodding and yelling at cattle, for example. This is caused, in part, by the frequent turnover of poorly compensated personnel, and by the tendency of humans to get frustrated and angry when a cow doesn't do what they want it to do. A simple pre-employment screening test, a questionnaire designed to measure empathy, would help management to hire the right sort of employees.

It's doubtful that this will happen unless cruelty is routinely discovered and begins both to carry a significant economic cost and to arouse social opprobrium. Grandin writes that managers aren't eager to be told that something being done under their supervision is cruel. Researchers who want to look into animal cruelty issues have great difficulty getting funding and even greater difficulty getting access to facilities.

It's mostly left up to states to protect farm animals. But many states are ignoring the overwhelming consumer support for better animal protection. Ironically, this is because the undercover videos have been so effective: They have resulted in expensive recalls, criminal convictions, slaughterhouse closings, lost sales, changes in practices, and changes in purchasing habits by big chains. Big Ag doesn't like this, and is very powerful in states where many slaughter facilities are located. Some of those states have passed "ag-gag" laws to criminalize the videotaping of animal cruelty and to punish those who distribute the resulting videos.[43] Iowa was first to adopt a law (2012) making it a crime to do undercover investigations of animal abuses. Utah, Kansas, North Dakota, and Montana quickly followed suit. Minnesota, Missouri, Tennessee, Florida, Nebraska, and even such progressive bastions as New York and California have contemplated similar laws.[44]

Trespassing, defamation, and libel laws already protect slaughterhouses from false claims. The real purpose of these new ag-gag laws is to keep us from knowing how our food is produced. Meat producers don't seem to realize that these laws can backfire and make Americans even more suspicious of mass-produced food.

Rather than making secret videos illegal, some ranchers would like to see mandatory video cameras record everything that goes on at slaughterhouses. Such cameras are widely used in Great Britain and could prevent unnecessary recalls. Nicolette Hahn Niman and her husband, Bill, have a grass-based organic beef ranch and got caught up in the recent Rancho Feeding Corporation recall of 8.7 million pounds of beef. In an op-ed for the *New York Times*, Nicolette told how they follow their own animals through every step of the slaughter process, working alongside USDA inspectors. Rancho handled both commodity and specialty beef and was the only slaughterhouse left in the Bay Area of California. Commodity cows were also processed at the plant, sometimes apparently without federal inspectors present. No cases of people getting sick were reported, but all the meat processed at the plant in 2013 was recalled. This was a devastating economic blow to the Nimans and other grass-fed/-finished beef ranchers in the Bay Area, who freeze much of their meat for sale later. Nicolette wrote that slaughtering is now the big bottleneck in the sustainable meat chain.[45]

Slaughterhouses are hard on humans as well. In 2012, the seventy-six thousand meat slaughterers and packers in the United States earned a median annual wage just over $25,000 a year.[46] Writer Eric Schlosser (*Fast Food Nation*) says four hundred cattle can zip through a slaughterhouse line in a single hour.[47] (JBS, the world's largest meat company, slaughters ninety thousand cows a day.) Lines move far too quickly at most slaughterhouses, which is cruel to both cows and workers. Processing facilities rarely have air-conditioning or heating. Workers have to stand nearly all the time. Sharp knives flash, and laceration is the most common injury.[48] Then there's the unquantifiable emotional toll on workers.[49]

Because everything is cheaper by the truckload, a problem for smaller grass-fed and specialty beef producers is the higher costs of slaughtering, packing, and distribution. Huge processing plants have to segregate specialty beef, which means idling and cleaning an entire production line. Small processors have to meet all the same USDA requirements, which means it might cost $500 per head instead of

$100, because only thirty head of cattle instead of five thousand head divide the cost. One solution is to sell customers part ownership of a live cow, which legally exempts the sale from some of the costs.[50]

For smaller farmers there is occasionally an alternative to fast-paced, consolidated, faraway disassembly plants. In Seattle, the "locavore" movement is strong, but smaller farmers have trouble getting local meat to consumers. Some found an answer in a nonprofit, cooperative-owned mobile meat-processing unit, a slaughterhouse on wheels. Accessorized with a USDA inspector, the truck goes from farm to farm. Similar mobile slaughtering units are springing up around the nation. Waste is left on the farm for composting, and cleaned, refrigerated, USDA-approved carcasses are taken to partnering cut-and-wrap facilities.[51] Such businesses may help smaller farmers survive and thus preserve farmland near cities. The USDA, however, has made it difficult for small slaughterhouses to get set up and certified. We're guessing here, but it might be because the USDA isn't allotted enough money to hire the necessary inspectors. And there's the NIMBY problem—no one wants to live downwind of an abattoir. But a mobile slaughterhouse, which would only be nearby intermittently, may help calm that fear. A little public education would also help: If consumers understand how cruel it is to ship cattle long distances, and how doing so affects the quality of meat, they might be more tolerant.

Taking the Fun Out of Cow Sex

A bull's genes greatly influence how much milk his daughters will produce and the quality of his progeny's meat. Excellent bulls are expensive to buy and maintain; it's usually more cost-effective to buy just their semen. A straw of semen from an excellent bull can cost as little as $25. A huge majority of dairy cows in North America are impregnated via artificial insemination. An increasing number of beef cows are also bred through artificial insemination, but most are still bred naturally with the willing cooperation of a bull.[52]

Research into the artificial insemination of cattle exploded in America in the 1940s. Gail's father, Paul Boyer, and his colleague, Henry Lardy, both then at the University of Wisconsin, took part in this research.[53] They helped develop ways to collect, evaluate, store, and ship cattle semen. Ways were also found to control ovulation in female cows, so the timing would be right. Skeptics thundered that using artificial insemination would result in milk unfit for human consumption and an "inferior, decadent, degenerative species."[54] On the contrary, artificial insemination helped to eliminate bad genes and to control venereal disease in cattle. It has, however, helped to turn dairy cows into the fragile specialists they are, and has led to some loss of genetic diversity.

Cows' attitudes toward artificial insemination haven't been studied, but it's obvious that bulls aren't into candlelit dinners and silk sheets. After a short training period, bulls seem eager to mount a fake cow and deposit into a fake vagina. Today, a steer might stand in for a cow, with a human standing by to introduce and hold the fake vagina at the appropriate moment. Learning about another method gave Denis the willies: A bull may be forced to ejaculate by being given an electric shock in his rectum. Massaging a bull's seminal vesicles and ampullae via his rectum is yet another approach.

Only girl cows make milk (duh), and the market for veal is shriveling, so researchers have found ways to determine the sex of sperm and sort it. One method is to mix sexing agent ($12) with the sperm in a straw of semen. The agent goads female sperm into greater activity so they win the race to the egg. It usually works. A machine to produce pre-sexed semen was developed with USDA funding. This method is about 90 percent accurate but results in a somewhat lower conception rate.

Because beef cattle are scattered across vast ranges, it's hard for a rancher to notice when the gals are in the mood (estrus). A cow collar developed in Scotland can send a text message to a farmer's cell phone if a female cow is making movements suggestive of being in estrus or going into labor. The gizmo uses the same technology as Wii video games.[55]

Embryo transplants, which began in the 1980s, are one way to get more calves from a choice beef or milk cow. A fine cow is treated with hormones to make her produce more than one egg. Five days later she's artificially inseminated with semen from a superior bull, and about a week later a flushing procedure removes about six to eight embryos from her uterus before their cells become specialized. Each of these tiny embryos can be implanted into a surrogate mom. Or the embryos can be stored in liquid nitrogen for future use. The resulting calves are genetically identical to one another, even though they were born from different mothers.[56] But this is just the start of human ingenuity.

With embryo transplants of milk cows, only half the DNA comes from the particular female the dairy farmer hopes to replicate. But with "fusion cell cloning," *all* the DNA (except mitochondrial DNA) comes from the most desirable cow.[57] Methods vary from lab to lab, but here's a rough idea: An unfertilized egg cell from a prize cow (cow A) is removed. Its nucleus (DNA) is gently removed and teamed up with a cell taken from the body (maybe from the ear) of a run-of-the-mill cow (cow B), from which the nucleus has been removed. The "fusing" of the nucleus and donor cell is done by giving them an electric shock. The resulting egg cell starts dividing and becomes an embryo that is then inserted into the uterus of a foster mother (cow C). The calf will be a clone of cow A. Both female and male cows can be nucleus donors.

Dolly the sheep, born in 1996, was the world's first fusion-cloned mammal (a tadpole was cloned in 1952). Dolly gave birth to six lambs created the natural way before developing terrible arthritis and lung cancer and being euthanized at the young age of six. The first cloned calf was born shortly thereafter.[58] The practice spread around the world and calves have now been cloned from all types of cattle and buffalo. Cloning can be done from live or dead cattle. Why use dead cows? Because you can't really tell how good a cow's meat is until it's cooked and served on a plate.

In 2008 the FDA declared cloned meat "virtually identical" to conventional meat and said that it's safe to eat cloned meat and safe to drink milk from cloned cows. The FDA therefore concluded that there was no need to label cloned products as such.[59] Nevertheless, the Consumer Federation of America, the Humane Society, and the Center for Food Safety, among others, are concerned about the quality of the science behind the FDA's decision, including the lack of peer reviewing. Surveys by the Pew Initiative on Food and Biotechnology found that 66 percent of Americans are "uncomfortable" with animal cloning. Forty-three percent don't believe such food is safe.[60] And 89 percent want cloned food labeled as such, reports Consumers Union.[61]

Meat from cloned cows is unlikely to turn up in significant amounts in grocery stores anytime soon. It is very expensive to clone a cow ($10,000 to $20,000 in 2008), so it is more likely that you will encounter meat from the *offspring* of cloned animals, which aren't technically clones. In August 2010, an FDA spokesperson said that because of a voluntary moratorium they were not aware of any cloned animal products already in the food supply.[62] If that statement was intended to encompass the calves of cloned animals, it seems improbable.[63] Milk from cloned cows has been sold for years.[64]

Late in 2010, the European Union placed a five-year ban on cloning animals for food production, but they did so mostly out of concern for the welfare of the animals. Far more deformities, miscarriages, and premature deaths occur in cloned pregnancies in cows. Because many cloned calves grow much larger as fetuses than normal, they require cesarean births. Cow clones are also prone to have malformations in their urogenital tracts, livers, and brains, and to be more vulnerable to infections.[65] Only about 6 percent of cloned embryos put in "birth-mother cows" result in healthy, long-lived offspring.[66]

The life span and other problems of clones appear to be influenced by epigenetic factors that better cloning techniques might be able to overcome.[67] The FDA insists that cloning techniques are improving. The agency's focus, however, is on whether cloned meat is safe for humans to eat, not on animal welfare. The offspring of clones (produced through sexual reproduction) don't seem to have the problems of their clone parent.[68]

At present, cloning is largely unregulated. We believe cloned cow products should be banned from human consumption unless they are humanely produced as well as safe to eat. We also support labeling genetically modified food as such, even if the FDA is correct that it's safe to eat. Consumers may care about the welfare of animals, the dominance of agribusiness oligopolies, and myriad other issues besides food safety.

Visit with a Bull Breeder

At this point it might be good to recall that nature can treat cows as roughly as humans do. To round out our research, we spent a day with Tom Elliott, former owner of the N Bar ranch in Montana, who bred the most influential Angus bull in twentieth-century America.

We met in Bend, Oregon, and sat at a picnic table under an apple tree to talk. Tom dresses like a cattleman: cowboy hat, boots, and a string tie with a silver slide depicting his most famous bull: N Bar Emulation EXT. We asked how he produced such a magnificent animal.

Turns out he did it just like Mother Nature would. When Tom first took over his grandfather's forty-eight-thousand-acre ranch, he tried all the usual breeding methods: boy-meets-girl, artificial insemination, and frozen embryos. Only then did he turn to natural selection.

Although as a young man Tom had had a lot of experience as a cowboy and ranch hand, his background was unusual for a bull breeder: He was a humanities major at Colorado State and earned an MBA from Duke. But he also has a deep interest in population genetics, conservation biology, quantum physics, and complexity science. Putting it all together, he decided to breed cattle the Darwinian way. He turned them out on the range and saw which ones perished and which ones survived and had qualities like calving ease, efficiency on

Tom Elliott, bull breeder extraordinaire. His outstanding bull Emulation EXT is depicted on his tie slide. Photo: Gail Boyer Hayes.

forage, a good balance of growth and carcass quality, and superior weaning weights. To get the best price when a cow is sold to a feedlot, most cattle breeders focus tightly on weaning weights and yearling weights. But this leads to unintended consequences. For example, breeding for a bigger size often results in reduced fertility.

Farm-raised cattle couldn't survive in most natural environments, Tom says. He thinks this is sad and wrong. So his herd was subjected to severe environmental pressures and heavily culled, a population genetics approach. A good herd finally in hand, with no known bad genes, he began "line breeding"—breeding prize fathers with outstanding daughters to produce superior bulls that can then be bred. Eventually the N Bar was selling four hundred bulls a year, and his most outstanding bulls produced an additional million dollars' worth of semen.[69] Tom admits that when so many calves come from one family it can tend toward a monoculture. But he argues, first, that his bulls were not bred for single traits but for their well-rounded ability to cope with everything nature could throw at them and, second, that half the genetic input in their offspring comes from the female, which helps maintain variety.

Cosmopolitan Cows

Dairy farms in Europe have been automated for two decades, but the practice is just beginning in the States. As of the start of 2014, Washington State had only three automated dairies. We wondered whether robots were kind to cows.

The morning was dark and cold when we arrived at the newly automated farm of Jim and Janie Austin. The first thing we saw was a well-maintained barn built in 1920, painted red with white trim. Jim, a man of seventy years who looks at least ten years younger, is one of those instantly likable people who win your trust. After a heart attack a few years ago, he sought ways to reduce his workload and acquire more free time. The farm wouldn't economically support another worker. Jim considered robots, but they required upfront capital. Ironically, in the eyes of those who think of organic farming and high technology as opposites, he was able to afford to mechanize by switching to organic production and selling his milk to the Organic Valley co-op. The Austins hadn't been using pesticides and herbicides anyway, so the transition from conventional to organic was easy. Jim was able to install two DeLaval milking robots, a manure scraper that cleans out the freestyle barn, and big carwash-like grooming brushes that clean off the cows.

The Austin's 130 cows now make their own decisions as to when, and how often, to be milked. The "milking parlor" is a clean white room. Two robotic machines are installed side by side on the far wall from the entrance. Day and night, cows walk up to the machines from a barn attached to the far side of the milking-room wall. They choose to be milked every eight to ten hours. Each cow's ear tag identifies her to the machines. A smart gate denies her entrance if she isn't ready to be milked.

During the half hour or so that we were in the room, there were always two lines of Holsteins and Jerseys patiently waiting to be milked. A gap in the wall between the machines allows the cows, one by one, to look over the humans (and the farm dog) in the parlor. When a cow is being milked, only her legs, udder, and the top of her back are visible from the parlor. The robots again read cows' ear tags, so they know exactly which lady is being milked. Red laser beams locate each of her tender-looking teats, wash and disinfect each one, and guide suction cups to fit over them. All the farmer

Automation, if well done, can increase cow comfort. Cows drink in front of a grooming brush on the Austin dairy farm. In the foreground is the automatic manure scraper. Photo: Gail Boyer Hayes.

has to do is to stay out of the way of the moving robotic arms. We learned that bull semen is already available that tends to produce dairy cows with straight, well-spaced, robot-friendly teats! By the third night after the machines were installed, the cows could be left on their own.

In the wall next to each robot is a computer with screens that record (from each individual teat) the amount of milk or colostrum obtained, the number of somatic cells in the milk, any blood in the milk, and the temperature and conductivity of the milk. (The amount of lactic acid in milk affects its conductivity; more acid means the cow has an infection. Measuring conductivity makes it easier to catch problems like mastitis about three days earlier than seen in somatic cell counts, and start treatment before an infection

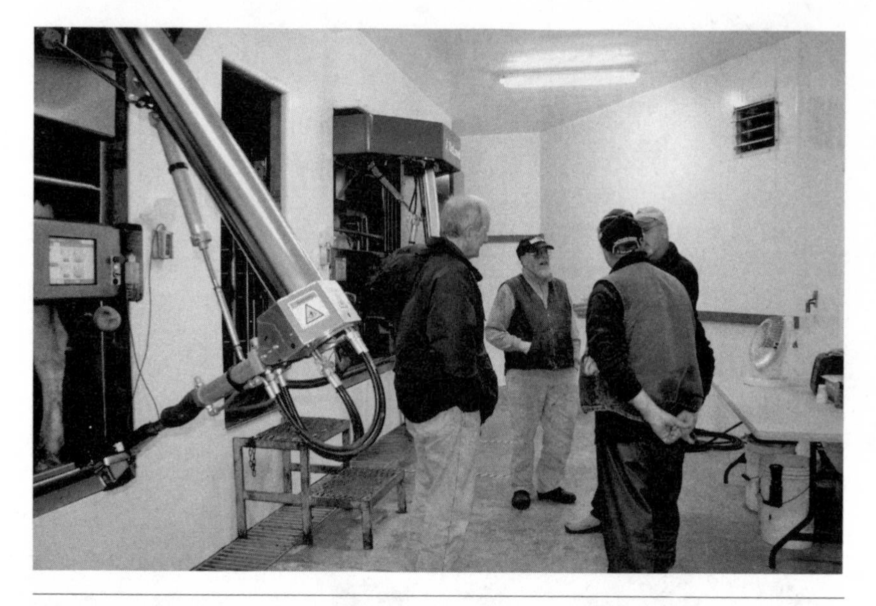

The two-robot milking parlor at the Austin farm has generated much interest among local farmers. Photo: Gail Boyer Hayes.

gets out of hand.) Jim is already seeing less mastitis, and milk production is up 11 percent.

The robots have replaced a farmworker (who quickly got a job on a nearby farm) as well as giving Jim an easier workday. The money formerly paid in wages, Medicare, and Social Security is about equal to the amount Jim pays to lease the machines. After seven years he has an option to buy the machines for 10 percent of their original cost. Robots even adjust cows' grain rations individually, so each cow gets exactly the diet she needs for where she is in her lactation cycle. The Austins now pay a bit more for grain.

Minnesotans Nels and Matt Goblirsch, third-generation dairy farmers, are other early adapters of robotic milking machines. Their 240 Holsteins also enjoy waterbeds. Nels told reporter Jenny Kirk of the *Marshall Independent* that they average three and a half milkings a day. Goblirsch cows now live longer—it's not unusual for a cow to

Jim Austin, a robotic arm, and a Holstein being milked. Both cows and farmer seem happy with the new automation.
Photo: Gail Boyer Hayes.

live ten to twelve years. They also produce more milk. Whatever their ladies really think about the new facility, the Goblirsches are clearly happier: They say the time they save can be used to spend more time on cow care.[70]

The success of robotic milking depends not just on the equipment but on farm management and other variables. The detection of mastitis is often where some systems fail. If well done, it is claimed that automatic milking systems can lead to a 12 percent increase in milk production and an 18 percent decrease in human labor.[71] The substitution of robots for workers across the breadth of the American economy raises troubling issues, but that is a topic for another book.

Robotic milking machines appear to offer the possibility of improving cow health and longevity. As far as we could tell, the cows seemed as happy to be milked by robots as by the humans we'd seen working with them on many other farms. It would be nice to think that adult cows like having humans around and enjoy being handled by them. But the grain they get as a reward while being milked, and their ability to choose for themselves when to relieve the pressure in their udders probably rate much higher in their priorities. And farmers can probably relate better to their cows when they have time to look at more than the beasts' back ends.

The unchanging rhythms of bovine existence can suppress human jitters due to the accelerating pace of modern life. Being near a friendly cow is calming. As the world spins its way through space, cows settle down and chew cud.

Sometimes cows mellow out even more. In the Okanagan Valley of British Columbia, a wine-producing region, some beef producers give each cow a liter of mixed red wines every day for sixty days, with the goal of producing flavored, better-textured meat. After taking a swig, the cows moo back and forth to each other more, are more docile, and seem more relaxed.[72] A winemaker in the south of France, Jean-Charles Tastavy, raised the ante, partnering with farmer Claude Chaballier to give Chaballier's cows two liters of locally produced wine every day. The liquid spirits lift the animals' spirits and raise the price the farmer gets for his meat.[73]

Following the lead of state-of-the-art dairies in California and the Midwest, in 2012 the third-generation Van Loon dairy in Oregon bought rubber waterbeds for three hundred cows. The Van Loons say the beds cut down on diesel fuel, medication, and bedding costs and that the cows are less stressed and more comfortable.[74] They report fewer sores and infections on cows enjoying the waterbeds.[75]

From that we've read and observed, *coddled* cows confined indoors display no more signs of stress than pastured cows. Yet it makes us uncomfortable to think of animals as big as cows being kept indoors all their lives. A popular video of Holsteins racing into a pasture when freed from their barn for the first time in the spring, kicking up their heels, big udders jiggling, is evidence that cows like being outdoors.[76] Freedom from acute stress isn't the equivalent of joy.

Moreover, the ecological costs of growing endless trainloads of monoculture corn, transporting it vast distances to dairies, pumping cows full of prophylactic antibiotics, and disposing of mountains of concentrated cow waste outweigh any ostensible benefits of CAFO dairies. Confining vast numbers of cows inside giant barns for their entire lives is wrong, even in the unusual cases when those cows are otherwise treated humanely. Consumers understand that intuitively. That's why CAFO milk is sold in cartons covered with pictures of cows in pastures.

Cows in Paradise: Pu'u O Hoku Ranch

Trailing a plume of red dust, we crested a hill, stopped, and stared down a forty-five-degree angle at a track studded with ottoman-sized volcanic boulders. Our driver hopped out of the ATV to unlock the gate separating us from the chasm. When she was again behind the wheel, we inched forward, the front of the ATV pointing down. We clung to the roll bar, and as the driver got out again to relock the gate we exchanged an are-we-crazy-to-do-this look.

It was February 2013 and we were on the Pu'u O Hoku Ranch, which consumes the best bite of the east end of the Hawaiian island of Molokai. Our driver was Jann Roney, manager of the ranch, and we were on a quest to locate the herd of beef cattle. A grandmother of nine, Jann has the nonstop energy of a toddler, the strong, tanned build of the Marlboro man, and the long, wavy (white) hair of a hula dancer. She comes from a family of seven generations of farmer-ranchers. Although she had intended to retire after her previous job, the possibilities of the sixteen-thousand-acre Pu'u O Hoku ranch so intrigued her that she signed on to help the owner make the ranch self-supporting, sustainable, environmentally sound, and an asset to the island. The ranch is now a sanctuary for native Hawaiian plants and animals, such as the endangered nene goose and monk seals. Its cattle produce beef that is grass-fed, grass-finished, and certified organic and biodynamic. As Jann says, "If we can't become self-reliant on an island like Molokai, we can't do it anywhere." Ironically, 95 percent of the beef eaten in Hawaii is shipped from the U.S. mainland, while forty thousand live cattle are annually shipped in the other direction. Interest in organic and local food is growing rapidly, even in Hawaii, but drought has hit the state, limiting supplies of grass.[77]

So it's a daunting challenge Jann has undertaken. Fortunately, the ranch is owned by a woman of means who is as determined and persistent as Jann is in wanting to meet their commendable goals. Both women are realists as well as idealists: Drought happens, cattle must be slaughtered, exploding deer populations must be controlled, squatters must be persuaded to move off ranch property.

Jann warned us that the cattle would be a bit thinner than we might expect because the first rains didn't fall until December 22. The grass on either side of the track was what we mainlanders call spring-green and hubcap-high. Jann said that with more rain it would now be a foot or two higher than a yardstick. Drought aside, this was paradise. When we dared to look, the Pacific sparkled in the distance, cliffs soared, and warm breezes carried bird chirp and a whiff of sea and flowers. Maui lay just to the south. The Pu'u O Hoku also leases land on the island of Oahu, to the north, where their yearling cattle nibble away under giant windmills. Barges are used to transport cows between islands.

Cattle on the Pu'u O Hoku just before they were rotated to a new pasture where they were up to their bellies in grass. Maui is in the distance. Photo: Gail Boyer Hayes.

During our long ride to find the cattle, while the ATV chewed and bounced its way over gullies and boulders, Jann filled us in on ranch history. Gail tried to take notes; the notes look like the tracings of a seismograph during a nine-pointer.

The ranch encompasses some of the most dramatic and beautiful scenery in Hawaii. Unfortunately, the land is also paradise for deer. The deer are believed to have been a gift to King Kamehameha V in 1867.[78] Aside from humans, there are no deer predators on Molokai. These delicate and fertile speckled Indian (aka chital or axis) deer have a voracious appetite for the same grasses cows prefer. Unlike cows, they can leap fences and nosh wherever they like, even in pastures that are supposed to be recovering from grazing. The deer do such environmental and crop damage that it is illegal to "keep" or transport them among Hawaiian islands.[79]

We asked, Why not shoot some deer with contraceptive darts or bullets to control their numbers? Well, locals have deep pride in their history on this thirty-eight-mile-long, salmon-shaped island. The product of two volcanoes, Molokai is now perhaps the most Hawaiian of the non–privately owned Hawaiian islands. Unemployment is high, household bank accounts are low, and locals have a taste for venison. If it weren't for the deer, the ranch's cattle operation would have a much better chance of making a profit off its fine beef, all of which is pre-sold. Not surprisingly, Jann is thinking of making and marketing both beef and venison jerky. Organic, of course.

For many reasons, it is important that ranch and locals coexist with aloha. One reason is that a hill on the ranch contains a special grove of kukui nut (candlenut) trees, which are now the Hawaii state tree. Kukui nuts were used in olden times as lights: The nuts were strung on the central rib of a palm leaf, a nut on the end was lit, and the nuts would burn one by one. This created a sort of clock as well as providing light. In Hawaii, kukui nut trees usually grow in gulches. What were they doing on a Pu'u O Hoku hill? Molokai singer Lono told us that this grove was planted to honor one of the greatest kahunas of ancient Hawaii, Lanikaula. The teacher, prophet, and sorcerer taught powerful magic and chants. The chants kept away attacking warships in old times and are said to have kept an unwelcome cruise ship from landing in modern times. Molokai locals feel strongly about protecting their island from becoming another Hawaiian Disneyland overrun by tourists. The ranch treats the grove with appropriate respect.

Until we rounded a hill a long distance from the gate, we didn't see a single cow. The forty or so ladies (the ranch does not keep steers) already had their heads lifted and were staring our way as we approached. They came in ebony, ivory, and russet. Aside from one cow (more about her later), these animals were not pets. As we pulled up close and stopped, Jann asked us to stay in the ATV.

Endangered nene geese with two goslings on the Pu'u O Hoku Ranch. Pastureland can provide habitat for endangered species along with cows. Photo: Gail Boyer Hayes.

Were it not for the deer, the ranch could sustainably support 450 cow/calf pairs. A previous owner had in excess of a thousand pairs. Combined with the rapidly expanding deer herd, this resulted in considerable damage to the land. The herd before us contained about thirty animals, roughly a third of the current herd of eighty-eight.

Pu'u O Hoku cows spend the first nine months of their lives, until they are weaned, with their moms. When they reach about five hundred pounds they spend another nine months out in open country with the herd, grazing and building strong frames. A final three months is needed to put on meat.

After just sitting and enjoying being with the cows for a bit, we drove toward the next fence. The cows knew the drill and followed. The ranch uses rotational grazing. Grass is all the cows eat, and grass always looks greener on the other side of the fence. California sour grass grows here as well as native grasses. The cows didn't much like sour grass at first. Then a happy surprise: The sour grass was mowed for compost and dried for two or three days. Cows gobbled it up.

Aged sour grass, a bovine treat—who'd have guessed? Maybe it's the bovine equivalent of sauerkraut.

Rather than sperm straws, this ranch keeps its own bulls and leaves procreation up to them. The cows here are mostly an Angus/Devon/Brangus mix, some with a touch of Hereford. (Brangus are a Brahman/Angus mix resistant to heat and humidity.)

Although the eastern end of Molokai is less vulnerable to drought than the west end, climate disruption is under way everywhere in the Pacific Islands. Temperatures are rising, rainfall decreasing, groundwater reserves shrinking, weather is more intense, and the ocean's chemistry and temperature are changing.[80] This ranch uses no irrigation to grow grass. Eleven of the past fourteen years have been drought years, 2012 being the worst. What further changes climate disruption will bring are uncertain. Downpours in this steep, volcano-formed land leave deep red gullies that look like wounds. Soil is precious.

A goal of the ranch, beyond succeeding financially, is to share the lessons of its success with other islanders. Molokai was once called the breadbasket of the islands. Today, fields of pineapple have been replaced by Monsanto GMO corn, grown for its seed on 1,850 leased acres near the island's center. Monsanto jobs are welcome on an island desperate for employment and income, but the company is also loathed by many islanders.

The ranch has only one *paniolo* (cowboy) to help Jann with the cattle on Molokai. Another part-timer works on Oahu. Many *paniolos* are descendants of Mexican vaqueros, imported to help Hawaiians learn the ropes, so to speak, and to ride horses. Island music honors these men in songs sung in Spanish and Hawaiian.

When a cow reaches 1,000 to 1,200 pounds, it is sent to a small cooperative slaughterhouse where things run on relaxed Hawaiian time and each cow, Jann assured us, is properly stunned. Killing and butchering animals is no one's dream job (humans are the only animals known to feel empathy with their food), but it has to be someone's work if we are to be omnivores.

As we left the cows behind, we asked Jann about the one cow that had trotted right up to her to be petted. Jann said that the cow's name was Maggie, and that none of the other cows had names. She slowed the ATV to eye a calf-sized boulder. "That's new. Must of rolled down the hill."

We learned that shortly after Maggie's mother dropped her, another cow walked over Maggie, injuring her back. The newborn was unable to walk and an infection set in. By the time she was discovered by a cowboy, her flesh was covered with maggots. Hence—in a dose of *paniolo* humor—she was named Maggie.

The ATV hit a challenging stretch of track and Jann fell silent for a bit.

The *paniolo* took the calf to Jann, who cleansed and treated the wound, nursed the calf with bottles of milk, and helped her get on her feet and walking. She sent the *paniolo* and calf back to the pasture so the calf could reunite with her mother. The *paniolo* dropped Maggie off near the herd, assuming her mother would find her. Jann's experience is that maternal dedication is far from universal. Silently cursing the male *paniolo's* lack of intuition, Jann began to worry. Although it was late in the day, she set out on a horse to find the calf. It was dusk by the time she got to the herd, which was in grass so tall she could barely see the grown cows. Calves were completely hidden. Finding Maggie appeared impossible.

Without really thinking, Jann began to moo. She demonstrated; her moo is eerily good. Maggie immediately responded. No lactating cow was around. Night was imminent and the landscape was treacherous. But Pu'u O Hoku translates into English as "Hill of Stars." The Milky Way blazed in a way never seen by cows in lit barns. Jann slipped a rope around Maggie's neck and began to lead her toward the ranch house. Hours later they arrived at ranch headquarters and Maggie soon became a permanent fixture around the shop and cowboy quarters. Fed by bottle, for over a year she became a bit of a ranch mascot and traveled with Jann to local grade schools. This allowed kids to have a chance to feed Maggie a bottle, pet her, and learn a little about cattle. It was also good PR for the ranch.

Compassion, dedication, persistence, and an intuitive feel for handling cows are what distinguish real ranchers from cow-factory managers. Jann would never be mistaken, nor want to be mistaken, for a Sierra Clubber, but she has a hard-nosed understanding of what is needed for a ranch to survive that most Sierra Clubbers would admire. The Pu'u O Hoku's cows are more than just economic variables to be maximized. Jann and the ranch owner take the long view—they have a hundred-year plan for restoring the ranch.

Few of the ranchers and dairy families we met have the deep pockets needed to do all they'd like for their cows, but the happiest of them have found ways to treat their animals incomparably better than do the giant confined operations that produce the bulk of America's milk and beef. It's going to be those farmers with vision and compassion who will prosper as the weather grows more erratic, as Earth's living skin is abraded by a growing number of humans, and as Americans pay closer attention to the sources of our food.

Afterword

THE TAIL END OF AN ERA

Few of today's cows could survive in the wild without humans to feed and shelter them. Humans could survive without cows, but our lives would be the poorer for it. We benefit greatly from cows. In addition to meat and dairy products, the search to use every cow component has yielded a dazzling array of goods; even cow ear hair is put to good use in making artist's and calligrapher's brushes. A few years ago, Gail's mother became part cow—she had a worn-out mitral valve in her heart replaced with one made from a cow's pericardial tissue. The anonymous donor cow has a special place in our own hearts.

So we want to keep cows around. A herd containing no more cows than a farm has acreage nearby to grow food for—and to absorb the waste of—makes more sense than a mega-factory farm holding tens of thousands of cows. If properly managed, smaller herds help restore badly damaged landscapes.

Huge feedlots should be phased out. Feedlot cows are miserable, standing in their own waste while being dosed with growth hormones and antibiotics. Their concentrated, contaminated fecal material is so excessive that there is no practicable way to safely dispose of it. To grow enough grain to supply feedlots, farmers plant vulnerable monocultures that require unimaginable amounts of fertilizers, pesticides, herbicides, and water. The result is an alarming loss of topsoil and an infusion of toxins into our communities.

The end of feedlots will mean the end of cheap grocery story beef. Beef will become a special food, served less often or in smaller portions, a food to be savored. Organic dairy products may continue to cost somewhat more than factory-farm-produced dairy; they are worth more.

Smaller dairies, farms, and ranches don't necessarily treat cows better than large operations. But their owners are closer to their animals, more likely to attend to them personally, and more likely to see them as individuals, not production units. This might result in more humane treatment of cows.

Raising far fewer but better cared-for cows will also benefit people. Americans now consume more artery-hardening beef than the people of almost any other nation. We also consume far more dairy products and protein than many public health professionals recommend. By cutting our meat and dairy consumption in half—which will result in our consuming an amount that is still more than ample to meet our needs—we can afford to buy smaller, healthier portions from well-treated cows with natural bovine diets. In 2013 an average American ate 57.5 pounds of beef per year (down from 61.6 a decade earlier).[1] If we can pare that down to 26 pounds—half a pound a week—and limit it to the healthiest, most sustainable beef, we can reduce pollution, global warming, medical costs, animal cruelty, loss of soil, loss of biodiversity, and germs resistant to antibiotics, while increasing the amount of land and water available for other uses. It's a dozen-for-one sale.

In a nation obsessed with diets, you might think of this as the Thomas Jefferson diet. Our third president was a health-conscious man who preferred vegetables to meat. Jefferson ate meat not as a main course but only as a "condiment" to his vegetable courses.[2]

Consuming less cow is not pointless symbolism or a diversion from political action. In fact, it is a profoundly political act, a repudiation of the powerful bovine industrial complex with its Beltway lobbyists and Madison Avenue hucksters. Every movement begins with individual acts. Afterward, with luck and good organization, the majority can abruptly change direction. It happened with smoking. It happened with recycling. It happened with same-sex marriage. But the individual acts came first.

Thanks to books by visionaries like Lester Brown (*Outgrowing the Earth*), Frances Moore Lappé (*Diet for a Small Planet*), Eric Schlosser (*Fast Food Nation*), Michael Pollan (*The Omnivore's Dilemma*), and to myriad local and national organizations supporting organic foods, locally grown foods, and vegetarian diets, Big Ag already knows we're out there. The industry pays close attention to what is selling and what is not. Farmers' markets have sprung up in nearly every city. Hundreds of thousands of families across the nation have signed up for Community Supported Agriculture. They pay in advance and get a box of vegetables weekly during the growing

Granddaughter Sheridan at a beginners' milking class. It's never too early to introduce children to cows. Photo: Lisa A. Hayes.

season—ultra-fresh veggies from "their" farm.[3] Some CSAs provide local, organic beef or dairy. Many restaurants and shops also feature locally grown and/or organic beef and dairy products. Thank those that do, and ask courteous questions of those that don't. That's how the campaign against smoking built momentum.

Young or old, omnivore or vegetarian, we've all had our lives greatly enhanced by the large, obliging beasts our ancestors sculpted out of aurochs. Cows, on the other hand, have had their lives cruelly diminished in the process. New forms of mass production treat animals with about the same level of respect that an automobile manufacturer feels for a piece of sheet metal. Our cows are a reflection of who we are as a people. The image in the mirror has grown disturbing. Let's change that. Doing so could be the beginning of many good things.

Acknowledgments

With thanks to, and admiration for, the hardworking men and women pioneering sustainable approaches to dairy farming and ranching who gave so generously of their time and talent and without whom this book would not have been possible: Bryce Andrews, Jim Austin, Jon Bansen, Karen Bumann, Tom Elliott, Leah Hair, Tom Herlihy, Kevin Maas, Wendy Millet, Kelan Moynagh, Jann Roney, Arturo Sandoval, Kurt Timmermeister, and Bryan Ulring.

We are indebted to friends and family members who volunteered to read parts or all of this work for accuracy, grace, and effectiveness: Paul D. Boyer, Lyda Boyer, Anne Biklé, Howie Frumkin, Dwight Gee, Jay Gordon, Sal Gordon, Deborah Jensen, Jeanne Klein, Steve Malloch, Jeanne Mathews, Joseph Massucco, David R. Montgomery, Susan Ott, Joanne Silberner, Joe Winston, and Barbara Wright. Any mistakes are ours, not theirs. Kim Girton, graphics and publications manager at Food & Water Watch, customized the map of beef feedlots above the Ogallala Aquifer to suit our needs. Liz Salim supplied the photos from Wisconsin. Gail offers special thanks to Nancy Eliasen, Myra Meeker, Marsha Olch, Iona Stenhouse, and Sally Wagner, the Leschi Ladies Literary League, for their sustained moral support and suggestions during this long project.

This book would never have been possible without our skilled guides through the publishing process: Elizabeth Wales and Maria

Guarnaschelli. Their astute early comments improved the manuscript a hundred fold. Mitchell Kohles, Laura Goldin, Sophie Duvernoy, and Lauren Maturo kept everything moving along. Meticulous copyediting by Janet Byrne prevented what would have been several mortifying mistakes.

And thank you, cows. You've put up with us for ten thousand years, given us meat and milk, pulled our plows, and pulled your weight in helping to make Western civilization possible.

Notes

Introduction
SETTING OUT

1. *Scientific American*, "Cows Tend to Face North-South," Aug. 26, 2008, http://www.scientificamerican.com/podcast/episode.cfm?id=FB9B3CC3-E299-DCDD-838FD534037D00CB. It took Google Earth to bring this to scientists' attention. See also Bob Yirka, "Magnetic Cow Findings Cause Row among Google Earth Researchers," PhysOrg.com, Nov. 15, 2011, http://phys.org/news/2011-11-magnetic-cow-row-google-earth.html.
2. Gidon Eshel et al., "Land, Irrigation Water, Greenhouse Gas and Reactive Nitrogen Burdens of Meat, Eggs, and Dairy Production in the United States," *Proceedings of the National Academy of Sciences*, July 21, 2014.
3. By the time we finished writing this book (Jan. 2014), the national herd had shrunk to 87.83 million head, the smallest since 1951. This was because of drought; the numbers will likely go back up—unless readers of this book effect a change. Ron Plain, "Annual Cattle Inventory," University of Missouri, Feb. 3, 2014, http://agebb.missouri.edu/mkt/bull12c.htm.

One
BEWITCHED BY COWS

1. *The Classics: Greek and Latin*, vol. 5 (*The Latin Classics*), ed. Marion Mills Miller, trans. W. A. McDevitte (New York: Vincent Parke and Co., 1909), 38–48, http://www.shsu.edu/~his_ncp/Caesar.html, a Sam Houston State University webpage (accessed May 30, 2013).

2. David Brown, "Scientists Unravel Genome of the Cow," *Washington Post*, Apr. 24, 2009.
3. For cow lovers, there's a fun app for your iPhone or iPod Touch called "My Daily Cow," http://krankykids.com/apps/mydailycow/. It tells you about a different cow every day and costs $2.99, which goes toward supporting free school programs.
4. The International Livestock Institute (Nairobi, Kenya) points out a few breeds well worth saving, including the Kuri longhorn of the Sahel, which can both survive drought and swim like Michael Phelps. In the United States, the American Livestock Breeds Conservancy helped save Randall cattle, which do well on small forage-based farms. Since 1999, the SVF Foundation at Tufts University's Cummings School of Veterinary Medicine has been collecting embryos and semen straws from rare and heritage breeds of livestock, including cattle with unusual survival traits and abilities. The initials "SVF" stand for Swiss Village Farm.
5. Gregory Cochran and Henry Harpending, *The 10,000 Year Explosion: How Civilization Accelerated Human Evolution* (New York: Basic Books, 2012), chap. 6.
6. Katarzyna Czechowicz, "Polish Geneticists Want to Recreate the Extinct Auroch," Nov. 28, 2011, http://www.eduskrypt.pl/polish_geneticists_want_to_recreate_the_extinct_auroch-info-11682.html. Translated by Google.
7. Jared Diamond, *Guns, Germs, and Steel* (New York: W. W. Norton, 1997). Diamond does discuss how the lack of domesticated animals was a drag on development in parts of the world—see pp. 141, 157–75.
8. The arrival of cattle on the East Coast and the introduction of cattle by the missions is based on Charles Wayland Towne and Edward Norris Wentworth, *Cattle & Men* (Norman: University of Oklahoma Press, 1955), 119–20.
9. Terry G. Jordan, *North American Cattle Ranching Frontiers: Origins, Diffusion, and Differentiation* (Albuquerque: University of New Mexico Press, 1993), 149, 210.
10. Ibid., 10–11, 236–37, 239. On the Great Plains, settlers used a mixed system of cattle raising, giving more attention to their cows than the longhorn ranchers. Although it's widely believed that the Anglo-Texans and their cows conquered the semi-arid Plains, Jordan disproved this theory.
11. Adrienne Koch and William Peden, eds., *The Life and Selected Writings of Thomas Jefferson* (New York: Modern Library, 1944), 280 (from Jefferson's *Notes on the State of Virginia*).
12. John Wesley Powell, *Report on the Lands of the Arid Region of the United States, with a More Detailed Account of the Lands of Utah. With Maps* (Washington, D.C.: Government Printing Office, 1897): 21–22.

13. Bureau of Animal Industry, *Annual Report of Animal Industry for the Year 1885*. (Washington, D.C.: Government Printing Office, 1886), 173, http://archive.org/details/annualreportbur06unkngoog.
14. Towne and Wentworth, *Cattle & Men*, 202.
15. Drew Conroy, *Oxen: A Teamster's Guide to Raising, Training, Driving, and Showing* (North Adams, MA: Storey Publishing, 2007), 10.
16. LeRoy R. Hafen and Francis Marion Young, *Fort Laramie and the Pageant of the West, 1834–1890* (Lincoln: University of Nebraska Press, 1984), 256–60.
17. Conroy, *Oxen*, 258.
18. Jennifer Moskowitz, "The Cultural Myth of the Cowboy, or How the West Was Won," *Americana*, Spring 2001.
19. M. R. Montgomery, *A Cow's Life: The Surprising History of Cattle, and How the Black Angus Came to Be Home on the Range* (New York: Bloomsbury, 2009), 66.
20. Montgomery, *A Cow's Life*, 66.
21. James P. Byrnes, David C. Miller, and William D. Schafer, "Gender Differences in Risk Taking: A Meta-Analysis," *Psychological Bulletin* 125, no. 3 (May 1999): 367–83. See also B. Pawlowski, Rajinder Atwal, and R. I. M. Dunbar, "Sex Differences in Everyday Risk-Taking Behavior in Humans," *Evolutionary Psychology* 6, no. 1 (2008): 29–42.
22. David T. Courtwright, *Violent Land: Single Men and Social Disorder from the Frontier to the Inner City* (Cambridge, MA: Harvard University Press, 1998), 3: "These troublesome elements—the surplus of young men, wide-spread bachelorhood, sensitivity about honor, racial hostility, heavy drinking, religious indifference, group indulgence in vice, ubiquitous armament, and inadequate law enforcement—were concentrated on the frontier. . . . [T]he frontier was . . . the most youthful and masculine region of the country and, consequently, the one most prone to violence and disorder."
23. Richard E. Nisbett and Dov Cohen, *Culture of Honor: The Psychology of Violence in the South* (Boulder, CO: Westview Press, 1966). For a different view, see Rebekah Chu, Craig Rivera, and Colin Loftin, "Herding and Homicide, An Examination of the Nisbett-Reaves Hypothesis," *Social Forces* 78, no. 3 (Mar. 2000): 971–87. Their take is that in the rural counties of the South, when poverty rates are controlled for, white male homicide rates were the opposite of what Nisbett and Cohen predicted. They argue that poverty, not herding, predisposes one to murder.
24. Paul F. Starrs, *Let the Cowboy Ride* (Baltimore: Johns Hopkins University Press, 1998), 256: "[T]he control of images and history by ranchers and westerners is . . . a sophisticated attempt to resist the weighty impositions

of federalism of government control from afar. . . . This they have done with astonishing effectiveness."

25. Russell W. Belk, "Shoes and Self," *Advances in Consumer Research* 30 (2003), http://www.acrwebsite.org/search/view-conference-proceedings.aspx?Id=8730.
26. Josie Glausiusz, "How Green Are Your Jeans?" *OnEarth* (NRDC), Dec.1, 2008, http://www.onearth.org/article/how-green-are-your-jeans.
27. Dan M. Kahan, "Fixing the Communications Failure," *Nature* 463 (2010): 296–97. Kahan and his colleagues did a revealing study in which they had actors present arguments for and against vaccinating girls to protect them from human papillomavirus (HPV). When the "expert" in favor of vaccination had a beard, wore a denim shirt, and had liberal-sounding articles on his resume, and the "expert" against vaccinating wore a suit and had a conservative list of publications, people became very polarized. When the clothes and publications were switched, the polarization disappeared.
28. Michigan Department of Agriculture & Rural Development, "Calling for Cattle," http://www.michigan.gov/mdard/0,4610,7-125-2961_2971_34984-72785—,00.html.
29. Julie Karceski, "Cow Brain Protein May Hold Alternative Energy Promise," *SLAC Today*, Apr. 20, 2010, http://today.slac.stanford.edu/feature/2010/cow-brains.asp.
30. Randolph E. Schmid, "Cow Genome Unraveled in Bid to Improve Meat, Milk," *USA Today* (AP), Apr. 24, 2009.
31. Georgetown University "Cow Genome Research Provides Clues to Evolution," news release, Feb. 28, 2012, www.georgetownedu/story/1242665711585.html.
32. "GM Cow Milked for Human Growth Hormone," *NewScientist*, Jan. 8, 2005. Pampa Mansa produces so much human growth hormone in her milk that a small herd of cows like her could meet world demand for it at a cost far less than that of the human growth hormone now made by bacteria. A male, Pampero, was born to Pampa Mansa, and his sperm can create more cows that produce human growth hormone—via the more efficient and kinder method of natural reproduction.
33. Society of Chemical Industry, "Skimmed Milk—Straight from the Cow," *ScienceDaily*, May 28, 2007, http://www.sciencedaily.com/releases/2007/05/070528084649.htm. Marge is not alone in this ability. The Milk Genomics Initiative at Wageningen University in the Netherlands is creating a database that includes cows with the ability to produce variations in milk fat.
34. Louis-Marie Houdebine, "Production of Human Polyclonal Antibodies by Transgenic Animals," *Advances in Bioscience and Biotechnology* 2 (2011): 138–41.

35. Ali Morton Walsh et al., "Widespread Horizontal Transfer of Retrotransposons," *Proceedings of the National Academy of Sciences* 110, no. 3 (Apr. 2013). See also: Julian Swallow, "No Bull! Research Shows Cows Could Share Up to a Quarter of Their DNA with Reptiles," *The Advertiser* (Australia), Jan. 3, 2013.
36. Alok Jha, "First British Human-Animal Hybrid Embryos Created by Scientists," *The Guardian*, Apr. 1, 2008. There was no intent in this experiment of planting such cells in a human or bovine mother.

Two
WHOA! COWS' BIG BITE

1. Pierre J. Gerber et al., *Tackling Climate Change through Livestock—A Global Assessment of Emissions and Mitigation Opportunities*, Food and Agriculture Organization of the United Nations, Rome, Sept. 2013.
2. David Archer, *The Long Thaw: How Humans Are Changing the Next 100,000 Years of Earth's Climate* (Princeton, NJ: Princeton University Press, 2008), 11.
3. United Nations News Centre, "Rearing Cattle Produces More Greenhouse Gases Than Driving Cars, UN Report Warns," Nov. 29, 2006, http://www.un.org/apps/news/story.asp?newsID=20772&#.U4zMrvldXa4.
4. University of Alaska, Fairbanks, "Methane Releases from Arctic Shelf May Be Much Larger and Faster Than Anticipated," *ScienceDaily*, Mar. 5, 2010.
5. William J. Ripple et al., "Ruminants, Climate Change and Climate Policy," *Nature Climate Change* 4, nos. 2–5 (Jan. 2014).
6. Henning Steinfeld et al., *Livestock's Long Shadow: Environmental Issues and Options* (Rome: Food and Agriculture Organization of the United Nations, 2006), Executive Summary. Present methods of estimating methane emissions from cows need to be improved.
7. Scot M. Miller et al., "Anthropogenic Emissions of Methane in the United States," *Proceedings of the National Academy of Sciences* 110, no. 50 (Dec. 2013): 20018–22.
8. Melanie Ford et al., "Agriculture and the Carbon Pollution Reduction Scheme," Commonwealth of Australia, 2009, 12–13; Julian Cribb et al., "Sustaining Growth of the Northern Beef Industry," Northern Australia Land and Water Taskforce, Oct. 2009; Sam Brasch, "6 Ways to Fight the Menace of Cow Burps," *Modern Farmer*, Jan. 7, 2014.
9. Society of Chemical Industry, "'Burpless' Grass Cuts Methane Gas from Cattle, May Help Reduce Global Warming," *ScienceDaily*, May 8, 2008, http://www.sciencedaily.com/releases/2008/05/080506120859.htm.

10. University of Alberta, "Way to Cut Cattle Methane, Threat to Environment, by 25 Percent," *ScienceDaily*, May 9, 2009, http://www.sciencedaily.com/releases/2009/05/090507145752.htm.
11. Richard Eckard, "Reducing Methane from Dairy Cows: It's All in the Oil," *The Conversation* (Australia), Mar. 22, 2013.
12. Valerie Elliott, "Wind of Change on Farms as Cows Help to Save the Earth," *The Times* (London), Sept. 8, 2008.
13. Penn State, "Unusual Feed Supplement Could Ease Gassy Cows, Reduce Their Greenhouse Gas Emissions," *ScienceDaily*, Sept. 8, 2010, http://www.sciencedaily.com/releases/2010/09/100907113135.htm.
14. Steinfeld et al., *Livestock's Long Shadow.*
15. Julie Cart, "Texas Drought Has Farmers on the Ropes," *Los Angeles Times*, May 22, 2011.
16. EPA/Water Sense, "Water Supply in the U.S," www.epa.gov/watersense/pubs/supply.html.
17. Vaclav Smil, "Worldwide Transformation of Diets, Burdens of Meat Production and Opportunities for Novel Food Proteins," *Enzyme and Microbial Technology* 30 (2002): 309.
18. Alan B. Durning and Holly B. Brough, "Taking Stock: Animal Farming and the Environment," *Worldwatch Institute Paper* 103 (1991).
19. Lecture by Jonathan Foley at the University of Washington, Feb. 1, 2012.
20. Mike Brugger, "Water Use on Ohio Dairy Farms," Ohio State University Extension, 2007, pdf.communicationx.net/w/water-use-onohio-dairy-farms—online-w422.html.
21. James McWilliams, "Meat Makes the Planet Thirsty," *New York Times*, Mar. 8, 2014.
22. Beef USA, "Beef Industry Statistics," 2012, http://www.beefusa.org/beefindustrystatistics.aspx.
23. Donald A. Wilhite, "Dust Bowl," *Oklahoma Historical Society's Encyclopedia of Oklahoma History and Culture*," http://digital.library.okstate.edu/encyclopedia/entries/d/du011.html.
24. John Steinbeck, *The Grapes of Wrath* (New York, Penguin Classics, 2006); Timothy Egan, *The Worst Hard Time: The Untold Story of Those Who Survived the Great American Dust Bowl* (Boston: Houghton Mifflin, 2005).
25. NOAA, "North American Drought: A Paleo Perspective—20th Century Drought," http://www.ncdc.noaa.gov/paleo/drought/drght_history.html.
26. By grinding up fossilized bison molars and measuring various chemicals and carbon isotope ratios (different grasses have different ratios), scientists at the University of Washington and Stanford have learned where and when certain grasses grew in prehistoric times at nine locations on the Great

Plains. This gave them a much clearer idea of the changes in climate and vegetation over the past two hundred thousand years. At times the Plains have been so dry that northern Nebraska was filled with sand dunes. Vince Stricherz, "Bison Teeth Tell Tales of Climate, Vegetation," *UW Today* (University of Washington), Aug. 17, 2006.

27. K. J. Brown et al., "Fire Cycles in North American Interior Grasslands and Their Relation to Prairie Drought," *Proceedings of the National Academy of Sciences* 102, no. 25 (2005): 8865–70.
28. Connie A. Woodhouse et al., "Climate Change and Water in Southwestern North America Special Feature: A 1,200-Year Perspective of 21st Century Drought in Southwestern North America," *Proceedings of the National Academy of Sciences* 107 (2010): 21283–88.
29. Jane Braxton Little, "The Ogallala Aquifer: Saving a Vital U.S. Water Source," *Scientific American*, Special Edition, Mar. 2009.
30. The Ogallala was created ten million years ago, when the Rocky Mountains were still clawing their way to the surface and streams and rivers raced down their emerging eastern slopes, cutting valleys and channels. These cuts filled up with rocks and unconsolidated clay, silt, sand, and gravel. Water filled the spaces around the rocks, etc. But this process stopped long ago. Today what little water runs into the aquifer comes from rain and snow and a bit of recharge from surface water diversions. An impermeable layer of caliche (a hardened deposit of calcium carbonate) that lies between the water and the surface soil over much of the aquifer makes it even harder to recharge the Ogallala. Nebraska is lucky: The aquifer underlying most of that state recharges more quickly, is deep, and global warming might actually increase rainfall there.
31. Sorghum Checkoff, United Sorghum Checkoff Program, First Quarter Report: Oct.–Dec. 2012, http://sorghumcheckoff.com/wp-content/uploads/2012/11/quarterly_report_octdec_FINAL.pdf.
32. Emilie T. Pinkham, "Comprehensive Groundwater-Management Scheme Can Prevent the Imminent Depletion of the Ogallala Aquifer, *Journal of Energy and Environmental Law* 3, no. 2 (Summer 2012): 268–79.
33. Charles Laurence, "US Farmers Fear the Return of the Dust Bowl," *The Telegraph*, Mar. 7, 2011.
34. Elliott Blackburn, "City Supplier Buys Pickens' Water Rights," *Lubbock Avalanche-Journal*, Apr. 8, 2011, http://lubbockonline.com/local-news/2011-04-08/city-supplier-buys-pickens-water-rights.
35. Food & Water Watch, "Factory Farm Map," 2007, www.factoryfarmmap.org.
36. Laurence, "US Farmers Fear the Return of the Dust Bowl."
37. University of Michigan, Global Change program, "Land Degradation," lec-

ture, http://www.globalchange.umich.edu/globalchange2/current/lectures/land_deg/land_deg.html.

38. David R. Montgomery, *Dirt: The Erosion of Civilizations* (Berkeley: University of California Press, 2007), 3–4.
39. Ibid., 163.
40. Richard Manning, "The Oil We Eat: Following the Food Chain Back to Iraq," *Harper's*, Feb. 2004.
41. As buffers with their diverse plantings are plowed under, migrating species that have relied on way stations for thousands of years are affected. Monarch butterflies are literally starving as they try to complete their annual journeys. Michael Wines, "Monarch Migration Plunges to Lowest Level in Decades," *New York Times*, Mar. 14, 2013.
42. A good summary of conservation-related provisions may be found by clicking on "round-up" in Cat Lazaroff's blog, "What the Farm Bill Means for Conservation," *Christian Science Monitor*, Feb. 19, 2014, http://www.resource-media.org/wp-content/uploads/2014/02/Farm-Bill-2014-Conservation-Round-Up.pdf.
43. We found less information available on Illinois—number two in the corn and soy derby—but did learn what the Illinois Environmental Council reported in 1990: "[A]pproximately 200 million tons of soil, an average of about 6.3 tons of soil per acre, are being eroded each year on 33 million acres in Illinois. At current erosion rates, 1.5 bushels of topsoil are lost for every bushel of corn produced in the state" (Wen Huang, "Reducing Soil Erosion: Status of Illinois' Goal by 2000," *Illinois Periodicals Online*, Mar. 1991). Another serious threat to farmland in Illinois is the expansion of towns and cities.
44. The difference in estimates by the USDA and the Iowa Daily Erosion Project (IDEP) is due in large part to the different ways in which "ephemeral gullies" are treated. Farmers often fill in gullies before planting crops, so that they can plant more corn or soy and use their giant machinery more efficiently. But when heavy rain comes, the water has to go somewhere and the gullies reform, taking soil and crops along with them. The IDEP approach is to monitor soil loss after storms, when and where it occurs, rather than look for a less meaningful "average."
45. Craig Cox, Andrew Hug, and Nils Bruzelius, *Losing Ground*, Environmental Working Group, 2011. Executive Summary.
46. Ibid., Making It Right section. See also Rod Swoboda, "Cropland Rental Trends Have an Impact on Rural Iowa," *Farm Progress*, Mar. 31, 2010.
47. J. Gordon Arbuckle Jr., "Rented Land in Iowa: Social and Environmental Dimensions," Iowa State University Extension, 2010. Like many states, Iowa recognizes a "Covenant of Good Husbandry" that applies to lessors, but

good husbandry is a vague term that doesn't place sustainability over productivity. See Drake University/Leopold Center, Agricultural Law Center, "Sustainable Farm Lease," for help in writing a sustainable lease, sustainablefarmlease.org.

48. Iowa State University/University Relations, "ISU Researcher: Iowa Has Lower-Quality Topsoil Than 50 Years Ago," Feb. 17, 2009, http://www.public.iastate.edu/~nscentral/news/2009/feb/veenstra.shtml.
49. David Pimentel et al., "Environmental and Economic Costs of Soil Erosion and Conservation Benefits," *Science* 267 (Feb. 1995). This estimate includes fertilizer to replace lost soil nutrients, restoring soil tilth, undoing flood damage, dredging silt out of rivers, etc.
50. Colin Sullivan, "Can Livestock Grazing Stop Desertification?" *Scientific American*, Mar. 5, 2013.
51. John Thackara, "Greener Pastures," *Seed*, Sept. 30, 2013.
52. Michael Tobis, "Allan Savory, Freeman Dyson, and Soil Sequestration," *Planet 3.0*, Mar. 17, 2013; Adam Merberg, "Cows Against Climate Change: The Dodgy Science Behind the TED Talk," *Inexact Change: Thoughts on Science, Politics, and Social Progress*, Mar. 11, 2013; James E. McWilliams, "All Sizzle and No Steak," *Slate*, Apr. 22, 2013; Chris Clark, "TED Talk Teaches Us to Disparage the Desert," KCET, *Social Focus*, Mar. 15, 2013.
53. Jamus Joseph et al., "Short Duration Grazing Research in Africa," *Rangelands*, Aug. 2002.
54. Lal Rattan, "Carbon Sequestration and Climate Change," conference presentation (2004): slide 20 et seq. http://presenter.cfaes.ohio-state.edu/link/Ratan_Lal_5-7-12_-_Flash_%28Large%29_-_20120507_03.37.06PM.html.
55. Kristin Ohlson, "Could Dirt Help Heal the Climate?" *Discover*, May 2011, 11–12.
56. North Carolina State University, "Scientists Find That Grasslands Can Act as 'Carbon Sinks,'" *ScienceDaily*, press release, Jan. 15, 2001.
57. Plains CO_2 Reduction Partnership, "EERC's Plains CO_2 Reduction Partnership and Ducks Unlimited Announce Carbon Credit Program," press release, Aug. 8, 2008, http://www.enn.com/press_releases/2597.
58. Charles C. Mann, "Our Good Earth," *National Geographic*, Sept. 2008, 80–107.
59. William I. Woods et al., eds., *Amazonian Dark Earths: Wim Sombroek's Vision* (New York: Springer, 2008); Paul Taylor, author and ed., *The Biochar Revolution: Transforming Agriculture and Environment* (Melbourne, Australia: Global Publishing Group, 2010).
60. Nathanael Johnson, "Just Add Compost: How to Turn Your Grassland Ranch into a Carbon Sink," *Grist*, Jan. 16, 2014, http://grist.org/climate-energy/just-add-compost-how-to-turn-your-grassland-ranch-into-a-carbon-sink/.

61. Jerry D. Glover and John P. Reganold, “Perennial Grains: Food Security for the Future,” *Issues in Science and Technology*,” Winter 2010, 43.
62. John Horowitz, Robert Ebel, and Kohei Ueda, “No-Till Farming is a Growing Practice,” USDA Economic Research Service, *Economic Information Bulletin* 70 (Nov. 2010).
63. J. D. Glover et al., “Increased Food and Ecosystem Security via Perennial Grains,” *Science* 328 (June 25, 2010): 1638–39.
64. Most plants are perennials, so why did farmers get stuck with annual food crops, which produce only once? This probably happened because our ancestors saved seeds from annuals to replant, picking seeds that came from the most productive plants. This selection process led to better annuals. Perennials don’t have to be replanted, so there was no urgent need to frequently gather seeds from them.
65. Sarah Whelchel and Elizabeth Popp Berman, “Paying for Perennialism: A Quest for Food and Funding,” *Issues in Science and Technology*, 28, no. 1 (Fall 2011): 63–76.
66. Hearing before the Committee on Agriculture, Nutrition, and Forestry, United States Senate, “Investing in Our Nation’s Future Through Agricultural Research,” Appendix 2: Public and Private Financing of Agricultural Research and Development, Government Printing Office, Mar. 7, 2007, 145: “Briefly, private R&D is commercially oriented. Companies, which must hold down costs, concentrate R&D funds on research that is likely to result in sales and profits, preferably on research that will lead to intellectual property that can be protected by patents. They are little interested in research that will benefit their competitors. For example, more than 40 percent of private agricultural R&D budgets is invested in product development, compared with less than 7 percent in public agricultural research.”
67. Glover and Reganold, “Perennial Grains: Food Security for the Future,” 44.
68. Philip Brasher, “Scientists Work on Perennial Crops to Cut Damage to Land,” *Des Moines Register*, May 7, 2011.
69. Philip Brasher, “Perennial Corn Holds Hope for Cutting Environmental Damage,” *Des Moines Register*, May 3, 2011.
70. National Priorities, “Cost of War in Afghanistan Since 2001,” https://www.national priorities.org/cost-of/war-in-afghanistan/.
71. Charles S. Wortmann and Charles A. Shapiro, “Composting Manure and Other Organic Materials,” University of Nebraska Lincoln Extension, Institute of Agriculture and Natural Resources, June 2012, http://www.ianrpubs.unl.edu/pages/publicationD.jsp?publicationId=567.
72. Frederick C. Michel Jr. et al., “Effects of Straw, Sawdust and Sand Bedding on Dairy Manure Composting,” North Carolina State University College of Agriculture and Life Sciences, 2005.

73. Ontario Ministry of Agriculture and Food, "Calculating Fertilizer Value of Compost Bedding Pack," Dec. 2008, http://www.omafra.gov.on.ca/english/livestock/dairy/facts/fertilizervalue.htm.
74. John E. Losey and Mace Vaughan, "The Economic Value of Ecological Services Provided by Insects," *BioScience* 56, no. 4 (2006): 311–23.
75. Katherine Harmon, "Dung Beetles Follow the Stars," *Scientific American*, Apr. 2013, 24.
76. Charles Darwin, *The Formation of Vegetable Mould, Through the Action of Worms, with Observations on Their Habits* (London: John Murray, 1882), 4.
77. Attributed to research done at Rothamsted Research, Harpenden, Hertfordshire, England.
78. Kelan gets many calls from people interested in starting a worm business. He suggests they "start small" before investing their retirement savings and think hard about how and to whom they will market their product. He warns them to be on the lookout for scams: operators who want to sell a $50,000 start-up package and promise to buy everything you produce. Such operators often go out of business right after raking in some money.
79. Rola M. Atiyeh et al., "Changes in Biochemical Properties of Cow Manure During Processing by Earthworms (*Eisenia andrei, Bouché*) and the Effects on Seedling Growth," *Pedobiologica* 44 (2000): 709–24.
80. L. Allison, H. Jack, and Eric B. Nelson, "Suppression of *Pythium* Damping Off with Compost and Vermicompost," report to the Organic Farming Research Foundation, Jan. 18, 2010.
81. Clive A. Edwards et al., "The Conversion of Organic Wastes into Vermicomposts and Vermicompost 'Teas' Which Promote Plant Growth and Suppress Pests and Diseases," *Hong Kong Agriculture and Fisheries Promotion Association Annual*, 2007, 33–40.
82. Anne Raver, "The Grass Is Greener at Harvard," *New York Times*, Sept. 23, 2009. Harvard has even set up its own website on the topic: http://www.campusservices.harvard.edu/buildings-facilities/landscaping/organic-landscaping.
83. Sue Neales, "Sustainable Livestock Vow Not Just Hot Air," *The Australian*, Mar. 27, 2012.

Three
HOOFPRINTS ON YOUR TABLECLOTH

1. Dick Teresi, "Lynn Margulis," *Discover*, Apr. 2011, 69.
2. Justin Gillis, "Climate Change Seen Posing Risk to Food Supplies," *New York Times*, Nov. 2, 2013.

3. Annika Carlsson-Kanyama, "Climate Change and Dietary Choices—How Can Emissions of Greenhouse Gases from Food Consumption Be Reduced?" *Food Policy* 23, nos. 3–4 (1998): 277–93.
4. David Pimentel and Marcia Pimentel, "Sustainability of Meat-Based and Plant-Based Diets and the Environment," *American Journal of Clinical Nutrition* 78, no. 3 (Sept. 2003): 6605–35.
5. Vaclav Smil, "Eating Meat: Evolution, Patterns, and Consequences," *Population and Development Review* 28, no. 4 (Dec. 2002): 599–639, 633 n. 15.
6. Nathan Fiala, "How Meat Contributes to Global Warming, *Scientific American*, Feb. 4, 2009.
7. Anthony J. McMichael et al., "Food, Livestock Production, Energy, Climate Change, and Health," *The Lancet* 370 (Oct. 2007): 1253–63. See also Daniele Fanelli, "Meat Is Murder on the Environment," *NewScientist.com*, July 18, 2007, http://www.newscientist.com/article/mg19526134.500-meat-is-murder-on-the-environment.html.
8. Elizabeth Rosenthal, "Sweden Looks to Diet to Cut Global Warming," *New York Times* (International), Oct. 23, 2009.
9. Christopher Weber and H. Scott Matthews, "Food-Miles and the Relative Climate Impacts of Food Choices in the United States," *Environmental Science and Technology* 42, no. 10 (2008): 3508–13.
10. Association for Pet Obesity Prevention, "Obesity Facts & Risks," 2013, http://www.petobesityprevention.org/pet-obesity-fact-risks/. Centers for Disease Control and Prevention (hereinafter CDC) FastStats Homepage, "Obesity and Overweight," 2011–2012, http://www.cdc.gov/nchs/fastats/obesity-overweight.htm.
11. Kelly S. Swanson et al., "Nutritional Sustainability of Pet Foods, *Advances in Nutrition* 4 (Mar. 2013): 141–50.
12. Association for Pet Obesity Prevention, "Obesity Facts & Risks," 2013.
13. CDC FastStats Homepage, "Obesity and Overweight," 2009–2010.
14. Lance Gibson and Garren Benson, "Origin, History, and Uses of Corn (*Zea mays*)," Iowa State University, Department of Agronomy, revised Jan. 2002, http://agron-www.agron.iastate.edu/Courses/agron212/Readings/Corn_history.htm.
15. Committee on Biosciences, Research in Agriculture, National Research Council of the National Academy of Sciences, *New Directions for Biosciences Research in Agriculture* (Washington, D.C.: The National Academies Press, 1985).
16. USDA Agricultural Research Service, "Corn: Taking Genetic Stock." Originally published in *Agricultural Research*, Jan. 2000. See also: USDA Agricultural Research Service, Maize Genetics Cooperation Stock Center, "Maize COOP Information," http://maizecoop.cropsci.uiuc.edu.

17. David Brown, "Scientists Have High Hopes for Corn Genome," *Washington Post*, Nov. 20, 2009.
18. The Maize Genetics Cooperation/Stock Center at the University of Illinois at Urbana-Champaign and the North Central Regional Plant Introduction Station at Iowa State University, "Maize COOP Information," http://maize coop.cropsci.uiuc.edu.
19. USDA National Agriculture Statistics Service, "U.S. Corn Acreage Up for the Fifth Straight Year," June 28, 2013, http://www.nass.usda.gov/Newsroom /2013/06_28_2013.asp.
20. Processors also turn field corn into degradable plastics, beauty products, bottled beverages, crayons, soap, and batteries. Field corn is used in a "slow adhesion" wallpaper that gives do-it-yourselfers time to reposition the paper they're handling; field corn helps keep frozen pizza crust crispy; field corn is found in gypsum and plasterboard; and field corn helps keep lollipops from dripping. USA Emergency Supply, "Corn—All About Grains," https://www .usaemergencysupply.com/information_center/all_about_grains/all_about _grains_corn.htm#.U8AubVaViu4.
21. According to the Corn Refiners Association, nearly 4,000 food items in a typical grocery store contain corn ingredients. And that number doesn't include all the sweet, whole-kernel corn that comes in cans, freezer bags, and on the cob, http://www.sdcorn.org/page/Education/sub/Cornasfood.
22. American Association for the Advancement of Science, "Diet And Disease in Cattle: High-Grain Feed May Promote Illness and Harmful Bacteria," *ScienceDaily*, press release, May 11, 2001.
23. Mike McGraw, "Building Bigger Cattle: An Industry Overdose," *Kansas City Star*, Dec. 2012.
24. PBS *Frontline*, Interview with Bill Haw, 2002, http://www.pbs.org/wgbh /pages/frontline/shows/meat/interviews/haw.html.
25. Jennifer Couzin, "Cattle Diet Linked to Bacterial Growth," *Science* 281 (Sept. 1998): 1578. Comment on Francisco Diez-Gonzalez et al., "Grain Feeding and the Dissemination of Acid-Resistant *Escherichia coli* from Cattle," in same issue, at 1666. See also C. J. Hovde et al., "Effect of Cattle Diet on *Escherichia coli* O157:H7 Acid Resistance," *Applied and Environmental Microbiology* 65, no. 7 (July 1999): 3233–35; Todd R. Callaway et al., "Diet, *Escherichia coli* O157:H7, and Cattle: A Review After 10 Years," *Current Issues in Molecular Biology* 11, no. 2 (2009): 67–79.
26. Vaclav Smil, *Enriching the Earth: Fritz Haber, Carl Bosch, and the Transformation of World Food Production* (Cambridge, MA: MIT Press, 2001).
27. Haber used his chemical wizardry in other ways to help the war effort—most notably by developing chemical weapons for trench warfare, in violation of international law. His chemist wife was so disturbed by this that she

committed suicide the morning Haber left to oversee deployment of chemical weapons on the eastern front. Haber himself took chemical warfare in stride. "Death is death," he famously commented.

28. Pioneer Hi-Bred International Inc., "Developing a Superior Maize Hybrid," http://www.pioneer.com/CMRoot/Pioneer/About_Global/news_media/media_library/articles/maize_hybrid.pdf.
29. Iowa State University Office of Biotechnology, "Hybrid Corn: A Case Study for Bioethics," updated 2003, http://www.bioethics.iastate.edu/classroom/hybridcorn.html.
30. Earl L. Butz, "The Family Farm: Shall We Freeze It in Place or Free It to Adjust?," in George Horwich and Gerald J. Lynch, eds., *Food, Policy, and Politics: A Perspective on Agriculture and Development* (Boulder: Westview Press, 1989).
31. USDA Economic Research Service, Topics/Crops, "Corn: Background," updated July 17, 2013.
32. Michael G. Siemens, "Managing Holstein Steers for Beef Production," University of Wisconsin Extension, Jan. 10, 1996, 4.
33. Whitney McFerron and Jeff Wilson, "U.S. Corn Supply Shrinking as Meat, Ethanol Demand Send Crop Price Higher," Bloomberg.com, Apr. 7, 2011, http://www.bloomberg.com/news/2011-04-08/u-s-corn-supply-shrinking-as-meat-ethanol-demand-send-crop-price-higher.html. See also Melissa C. Lott, "The U.S. Now Uses More Corn for Fuel Than for Feed," *Scientific American* Blogs, Oct. 7, 2011, http://blogs.scientificamerican.com/plugged-in/2011/10/07/the-u-s-now-uses-more-corn-for-fuel-than-for-feed/.

    Robin Young and Jeremy Hobson, "Ethanol: Off the Radar, but Bigger Than Ever," Here & Now, Public Radio International, Apr. 19, 2012, http://hereandnow.wbur.org/2012/04/ethanol-use-energy.

    Craig Cox and Andrew Hug, "Driving Under the Influence: Corn Ethanol and Energy Security," Environmental Working Group (EWG), June 2010, 3, http://www.ewg.org/news/news-releases/2010/06/14/driving-under-influence-huge-taxpayer-investment-ethanol-yields-paltry. This study by the Environmental Working Group estimated cumulative ethanol subsidies between 2005 and 2009 at $17 billion. If the subsidies are continued to 2015, EWG estimated that they would cost $53.6 billion.
34. Matthew L. Wald, "For First Time, E.P.A. Proposes Reducing Ethanol Requirement for Gas Mix," *New York Times,* Nov. 16, 2013.
35. David Pimentel and Tad W. Patzek, "Ethanol Production Using Corn, Switchgrass, and Wood; Biodiesel Production Using Soybean and Sunflower," *Natural Resources Research* 14, no. 1 (Mar. 2005).
36. David Lorenz and David Morris, "How Much Energy Does It Take to Make

a Gallon of Ethanol?" Institute for Local Self Reliance, Aug. 1995. See also Pimentel and Patzek, "Ethanol Production Using Corn, Switchgrass, and Wood"; Tom Philpott, "An Interview with David Pimentel," *Grist*, Dec. 9, 2006. One exception is the energy credit given for the DDG by-product. Researchers who find a net benefit from corn tend to credit DDG as a substitute for corn feed; critics say the low-fat, high-protein DDG is actually being substituted for soybean meal and award it a *much* smaller energy credit.

37. Pimentel and Patzek, "Ethanol Production Using Corn, Switchgrass, and Wood."
38. Timothy Searchinger et al., "Use of U.S. Croplands for Biofuels Increases Greenhouse Gases Through Emissions from Land Use Change," *Science* 319 (Feb. 29, 2008): 1138–1240.
39. David Biello, "Fertilizer Runoff Overwhelms Streams and Rivers—Creating Vast 'Dead Zones,'" *Scientific American*, Mar. 14, 2008.
40. Glover and Reganold, "Perennial Grains: Food Security for the Future," 43.
41. The known exceptions are minuscule: certain cave bacteria that get energy from radiation, and deep-sea creatures feasting on sulfide-oxidizing bacteria around hydrothermal vents.
42. Vaclav Smil, "Harvesting the Biosphere: The Human Impact," *Population and Development Review* 37, 4 (Dec. 2011): 613–36.
43. Edward O. Wilson, *The Diversity of Life* (Cambridge, MA: Belknap, 1992).
44. Global Footprint Network, "Footprint Basics—Overview," July 18, 2012, http://www.footprintnetwork.org/en/index.php/gfn/page/footprint_basics_overview/.
45. Smil, "Eating Meat."
46. Neil MacFarquhar, "Food Prices Hit Record Levels, Fueled by Uncertainty, U.N. Says," *New York Times*, Feb. 4, 2011. See also Economist Intelligence Unit, "Global Food Security Index 2012," *The Economist*, http://pages.eiu.com/rs/eiu2/images/EIU_DUPONT_Food_Index_July_2012.pdf.
47. USDA Foreign Agricultural Service, "Grain: World Markets and Trade," May 2010, 3; May 2014, 3.
48. Douglas A. McIntyre, "U.S. Becomes World's Bread Basket Again," 24/7 wallstreet.com, Feb. 24, 2012, http://247wallst.com/commodities-metals/2012/02/24/u-s-becomes-worlds-bread-basket-again/.
49. Stephen Pincock, "High Food Prices Driving Unrest: Study," *ABC Science*, Aug. 25, 2011, http://www.abc.net.au/science/articles/2011/08/18/3295848.htm.
50. Food and Agriculture Organization of the United Nations, "The State of Food Insecurity in the World," 2013, http://www.fao.org/docrep/018/i3434e/i3434e00.htm.

51. United Nations Department of Economic and Social Affairs, Population Division, Population Estimates and Projections Section, "World Population Prospects: The 2012 Revision," http://esa.un.org/wpp/. An interesting side note in the UN study: If China's population continues to decline slightly and Nigeria's continues to skyrocket, Nigeria will replace China by 2100 as the world's second most populous country. This is an example of the fallacy of taking projections out too far. On the other hand, the forecast that Nigeria's population could overtake the United States by 2050 seems plausible; it may depend mostly on American immigration policies. Also worth reading on this subject is Gordon Conway, *One Billion Hungry: Can We Feed the World?* (Ithaca: Cornell University Press, 2012). A stark collection of facts coupled with a relentlessly upbeat assessment of how things will improve, by Lester Brown, is *Full Planet: Empty Plates* (New York: W. W. Norton, 2012). A careful distillation of a lifetime's work on this crucial topic, Brown (like Conway) offers genuine "solutions" to the daunting problems discussed but understands how divorced they are from today's global distribution of political and economic power.
52. *The Economist*, "Kings of the Carnivores," Apr. 30, 2012.

Four
LAGOON BLUES

1. EPA Clean Water Act, National Enforcement Initiatives, updated Feb. 7, 2014, http://www2.epa.gov/enforcement/national-enforcement-initiatives. This figure includes cattle, pig, chicken, horse, turkey, duck, and sheep CAFOs.
2. EPA, "National Pollutant Discharge Elimination System Permit Regulation and Effluent Limitation Guidelines and Standards for Concentrated Animal Feeding Operations (CAFOs), Final Rule," *Federal Register* 68, no. 29 (Feb. 2003): 7176–80.
3. Robbin Marks, *Cesspools of Shame: How Factory Farm Lagoons and Sprayfields Threaten Environmental and Public Health* (Washington, D.C.: Natural Resources Defense Council and the Clean Water Network, 2001), 33–34.
4. Doug Gurian-Sherman, *CAFOs Uncovered: The Untold Costs of Confined Animal Feeding Operations* (Cambridge, MA: Union of Concerned Scientists, 2008), 2; JoAnn Burkholder et al., "Impacts of Waste from Concentrated Animal Feeding Operations on Water Quality," *Environmental Health Perspectives* 115, no. 2 (Nov. 2006): 308–12.
5. Smil, "Eating Meat," 620.
6. Stephanie Paige Ogburn, "Idaho: The CAFO State?" *High Country News*, Aug. 22, 2011.

7. Idaho Statutes, Title 22, Chapter 49: 22-4909A, http://www.legislature.idaho.gov/idstat/Title22/T22CH49SECT22-4909A.htm.
8. A good overview of the Idaho situation can be found in Scott Weaver, "Cow Country: The Rise of the CAFO in Idaho," *Boise Weekly*, Sept. 1, 2010.
9. C. J. Hadley, "Mr. Spud," *Range*, Summer 1998, http://www.rangemagazine.com/archives/stories/summer98/jr_simplot.htm. See also *The Economist*, Obituary, "Jack Simplot; Jack Simplot, Potato- and Memory-Chip Tycoon, Died on May 25th, Aged 99," June 12, 2008; Paul Rogers and Jennifer LaFleur, "Cash Cows, The Giveaway of the West," *San Jose Mercury News*, Nov. 7, 1999; EPA Newsroom, "EPA Orders Simplot Cattle Feeding Company to Change Stock Watering Practice at Grand View, ID, Feedlot to Protect the Snake River," news release from Region 10, June 11, 2010; EPA, "Agriculture/Animal Feeding Operations—Compliance & Enforcement Cases 2009 through Present," http://www.epa.gov/oecaagct/anafocom.html; "Precision Pumping Systems Helps Feedlot Feed Millions," http://www.indusoft.com/Portals/0/PDF/CaseStudies/062911-CS_PPS-B-ENLT-WB.pdf.
10. Consumers Union Southwest Regional Office, *Animal Factories: Pollution and Health Threats to Rural Texas*, May 2000, http://consumersunion.org/pdf/CAFOforweb.pdf.
11. Marc Ribaudo et al., "*Manure Management for Water Quality Costs to Animal Feeding Operations of Applying Manure Nutrients to Land*," USDA Economic Research Service, Agricultural Economic Report 824 (June 2003).
12. EPA National Pollutant Discharge System (NPDES), Animal Feeding Operations, "Concentrated Animal Feeding Operations Final Rulemaking—Fact Sheet," Oct. 20, 2008, http://www.epa.gov/npdes/pubs/cafo_final_rule2008_fs.pdf.
13. GAO, "Concentrated Animal Feeding Operations: EPA Needs More Information and a Clearly Defined Strategy to Protect Air and Water Quality from Pollutants of Concern," GAO-08-944, Sept. 4, 2008.
14. NRDC, "EPA, Environmental Groups Reach Settlement on Factory Farm Pollution Lawsuit," press release, May 26, 2010.
15. EPA, Clean Water Act: Agriculture-Related Enforcement Cases (2013), http://www.epa.gov/oecaagct/lcwaenf.html.
16. Perry Beeman, "EPA Gives Iowa Last Chance to Inspect Livestock," *Indystar*, Dec. 30, 2012.
17. See http://harrisranch.com/company_history.php.
18. Jay Ham, "Managing Livestock Ammonia: A Volatile, Promiscuous Fugitive in the Atmosphere (Rocky Mountain National Park)," Washington State University Extension, updated May 22, 2013.
19. Gurian-Sherman, *CAFOs Uncovered: The Untold Costs of Confined Animal Feeding Operations*, 60.

20. Marks, *Cesspools of Shame*, 18.
21. Sid Perkins, "ScienceShot: There's Cow in Your Smog," News.sciencemag.org, May 1, 2012.
22. Marks, *Cesspools of Shame*, 18.
23. Washington State University Extension, "Anaerobic Digesters and Biogas Safety," updated Apr. 2012, http://www.extension.org/pages/30311/anaerobic-digesters-and-biogas-safety.
24. EPA, Ozone and Your Patients' Health: Training for Health Care Providers, "Health Effects of Ozone in the General Population," updated Mar. 12, 2014, http://www.epa.gov/apti/ozonehealth/population.html.
25. Janet Raloff, "Rural Ozone Can Be Fed by Feed (as in Silage)," *ScienceNews*, Apr. 21, 2012.
26. D'Ann L. Williams et al., "Airborne Cow Allergen, Ammonia and Particulate Matter at Homes Vary with Distance to Industrial Scale Dairy Operations: An Exposure Assessment," *Environmental Health* 10, no. 72 (2011). See also: Phil Ferolito, "Large Dairies, Feed Lots Linked to Air Pollution," *Seattle Times*, Dec. 17, 2011.
27. Maria Cone, "State Dairy Farms Try to Clean Up Their Act," *Los Angeles Times*, Apr. 28, 1998.
28. Bettina Schiffer et al., "The Fate of Trenbolone Acetate and Melengestrol Acetate After Application as Growth Promoters in Cattle: Environmental Studies," *Environmental Health Perspectives* 109, no. 11 (Nov. 2001).
29. Janet Raloff, "Hormones: Here's the Beef," *Science News* 161, no. 1 (Jan. 5, 2002): 10. See also Edward F. Orlando et al., "Endocrine-Disrupting Effects of Cattle Feedlot Effluent on an Aquatic Sentinel Species, the Fathead Minnow," *Environmental Health Perspectives* 122, no. 3 (Mar. 2004): 353–58.
30. David Kirby, *Animal Factory: The Looming Threat of Industrial Pig, Dairy, and Poultry Farms to Humans and the Environment* (New York: St. Martin's Press, 2010), 326. See also Kathy Dobie, "One Woman Takes a Brave Stand Against Factory Farming," *O, the Oprah Magazine*, Nov. 2011.
31. Pew Commission on Industrial Farm Animal Production, "Putting Meat on the Table: Industrial Farm Animal Production in America," Pew Charitable Trusts and Johns Hopkins Bloomberg School of Public Health, 2008, http://www.ncifap.org/_images/PCIFAPFin.pdf.
32. Aleksey V. Zimin et al., "A Whole-Genome Assembly of the Domestic Cow, *Bos Taurus*," *Genome Biology* 10, no. 4 (2009), Article R42.
33. John H. Kirk, "Tuberculosis in Cattle," May 16, 2002, http://www.learningace.com/doc/627869/3a5f034f9122be53fe8d2af1c2ec909f/tuberculosis.
34. William Shulaw and Randall E. James, "Rabies Prevention in Livestock," Ohio State University, n.d., http://ohioline.osu.edu/vme-fact/0001.html.

35. National Research Council, Committee on Diagnosis and Control of Johne's Disease, *Diagnosis and Control of Johne's Disease* (Washington, D.C.: The National Academies Press, 2003), 104.
36. Mark Jerome Walther, *Six Modern Plagues* (Washington, D.C., Island Press, 2003), 68.
37. Michael Moss, "The Burger That Shattered Her Life," *New York Times*, Oct. 3, 2009. See also Thomas G. Boyce et al., "*Escherichia coli* O157:H7 and the Hemolytic-Uremic Syndrome," *New England Journal of Medicine* 333, no. 6 (Aug. 1995): 364–68; Madeleine Baran, "Minn. Woman, Cargill Settle *E. coli* Lawsuit," Minnesota Public Radio, May 12, 2010; Bill Marler, "*E. coli* Lawsuit to be Filed against Food Giant Cargill on Behalf of Stephanie Smith," Dec. 3, 2009, http://www.marlerblog.com/legal-cases/e-coli-lawsuit-to-be-filed-against-food-giant-cargill-on-behalf-of-stephanie-smith/#.U4eKWPldXa4.
38. Moss, "The Burger That Shattered Her Life."
39. Callaway et al., "Diet, *Escherichia coli* O157:H7, and Cattle: A Review After 10 Years."
40. Robert O. Elder et al., "Correlation of Enterohemorrhagic *Escherichia coli* O157 Prevalence in Feces, Hides, and Carcasses of Beef Cattle During Processing," *Proceedings of the National Academies of Sciences* 97, no. 7 (Mar. 2000). Each "lot" of the twenty-nine to thirty lots sampled consisted of groups of fed cattle from a single source.
41. Bill Marler, "*E. coli* Recall of 1,000,000 Pounds of Meat from Valley Meat Company," Aug. 6, 2010, http://www.marlerblog.com/case-news/e-coli-recall-of-1000000-pounds-of-meat-from-valley-meat.company/.
42. CDC, "Multistate Outbreak of *E. coli* O157:H7 Infections Linked to Eating Raw Refrigerated, Prepackaged Cookie Dough," June 30, 2009, http://www.cdc.gov/ecoli/2009/0630.html.
43. Dennis G. Maki, "Don't Eat the Spinach—Controlling Foodborne Infectious Disease," *New England Journal of Medicine* 355, no. 19 (Nov. 2006).
44. USDA Food Safety and Inspection Service, "Microbiological Testing Program for *E. coli* O157:H7 and Non-O157 Shiga Toxin-Producing *E. coli*: Individual Positive Results for Raw Ground Beef (RGB) and RBG Components," 2013, http://www.fsis.usda.gov/wps/portal/fsis/topics/data-collection-and-reports/microbiology/ec/testing-program-for-e-coli-o157h7-and-non-o157-stec.
45. Helen Shen, "US Beef Tests Cook Up a Storm," *Nature* 485 (May 2012): 558–59. See also Stephanie Storm, "Government to Increase *E. coli* Tests in Raw Beef," *New York Times*, May 31, 2012; Andrew Schneider, "USDA May Be Ready to Tackle 'Other' *E. coli* Strains," *AOL News*, Sept. 28, 2010, http://www

.marlerclark.com/media_relations/view/usda-may-be-ready-to-tackle-other-e.-coli-strains.

46. A new Shiga-toxic-producing strain, O104:H4, appeared in Germany in the spring of 2011, killing over fifty people—mostly from kidney failure. Mary Rothschild, "Another Clue to E. coli O104:H_4?" *Food Safety News*, Mar. 8, 2012, http://www.foodsafetynews.com/2012/03/another-clue-to-e-coli-o104h4/#.U4eTkvldXa4.
47. David Pierson, "Machine-Tenderized Beef May Require Tougher Labeling," *Los Angeles Times*, Oct. 21, 2013.
48. Stephanie Paige Ogburn, "Cattlemen Struggle Against Giant Meatpackers and Economic Squeezes," *High Country News*, Mar. 21, 2011.
49. Mike McGraw and Alan Bavley, "Beef's Raw Edges" (series), *Kansas City Star*, 2012, http://kansascity.com/2012/12/06/v-project_one/3951690/beefs-raw-edges.html; http://www.kansascity/2013/06/25/4313446/stars-series-on-beef-industry.html. These were original links, but the series couldn't be accessed on the newspaper's website as of July 5, 2014. However, as of that date the series could be accessed on the McClatchyDC website: http://www.mcclatchydc.com/2012/12/09/176738/big-beef-beefs-raw-edges.html; http://www.mcclatchydc.com/2012/12/11/176740/big-beef-industry-fights-back.html; http://www.mcclatchydc.com/2012/12/10/176739/drug-overuse-in-cattle-imperils.html.
50. Pierson, "Machine-Tenderized Beef May Require Tougher Labeling," *Los Angeles Times*.
51. CDC, "No Progress in *Salmonella* During Past 15 Years," press release, June 7, 2011.
52. FDA, "Archive for Recalls, Market Withdrawals, & Safety Alerts," updated May 28, 2014, http://www.fda.gov/Safety/recalls/default.htm. See also Mary Clare Jalonick, "Romaine Lettuce Recall in 23 States over *E. coli*: Details," *Huffington Post*, May 6, 2010, http://www.huffingtonpost.com/2010/05/06/lettuce-recall-e-coli-pos_n_566956.html.
53. CDC, "Multistate Outbreak of Human *Salmonella* Enteritidis Infections Linked to Alfalfa Sprouts and Spicy Sprouts," June 28, 2011, http://www.cdc.gov/salmonella/sprouts-enteritidis0611/062611/.
54. Zach Mallove, "How Did *E. coli* O145 Contaminate Lettuce? Part III," *Food Safety News*, May 14, 2010. See also Michael Cooley et al., "Incidence and Tracking of *Escherichia coli* O157:H7 in a Major Produce Production Region in California," *PLoS ONE* 2, no. 11 (2007).
55. ConsumerReports.org, "Dole Bagged Salads Recalled Due to Salmonella Risk," Apr. 16, 2012, http://www.consumerreports.org/cro/news/2012/04/dole-bagged-salads-recalled-due-to-salmonella-risk/index.htm.
56. Maki, "Don't Eat the Spinach—Controlling Foodborne Infectious Disease."

57. E. B. Solomon, S. Yaron, and K. R. Matthews, "Transmission of *Escherichia coli* O157:H7 from Contaminated Manure and Irrigation Water to Lettuce Plant Tissue and Its Subsequent Internalization," *Applied and Environmental Microbiology* 68, no. 1 (2002): 397–400. See also E. B. Solomon, C. J. Potenski, and K. R. Matthews, "Effect of Irrigation Method on Transmission to and Persistence of *Escherichia coli* on Lettuce," *Journal of Food Protection* 65, no. 4 (Apr. 2002): 673–76; and M. C. Erickson et al., "Surface and Internalized *Escherichia coli* O157:H7 on Field-Grown Spinach and Lettuce Treated with Spray-Contaminated Irrigation Water," *Journal of Food Protection* 73, no. 6 (June 2010): 1023–29.
58. K. M. Lee et al., "Inhibitory Effects of Broccoli Extract on *Escherichia coli* O157:H7 Quorum Sensing and In Vivo Virulence," *FEMS Microbiology Letters* 321, no. 1 (Aug. 2011): 67–74.
59. The Beef Site, Beef Disease Guide, "Ringworm in Cattle," http://www.thebeefsite.com/diseaseinfo/233/ringworm-in-cattle.
60. Hannah Furness, "Resistance to Antibiotics Could Bring 'The End of Modern Medicine as We Know It,' WHO Claim," *The Telegraph*, Mar. 16, 2012.
61. Sabrina Tavernise, "Antibiotic-Resistant Infections Lead to 23,000 Deaths a Year, C.D.C. Finds," *New York Times*, Sept. 17, 2013.
62. Caroline Smith DeWaal, Cindy Roberts, and Caitlin Catella, "Antibiotic Resistance in Foodborne Pathogens: Evidence of the Need for a Risk Management Strategy," Center for Science in the Public Interest, Jan. 2011, Preface, 1.
63. Columbia University, School of International and Public Affairs, "Scientific Analysis of the Preservation of Antibiotics for Medical Treatment Act of 2011," http://mpaenvironment.ei.columbia.edu/sitefiles/file/Summer%2011%20reports/Preservation%20of%20Antibiotics_%20Final%20Report.pdf; D. I. Andersson, "Persistence of Antibiotic Resistant Bacteria," *Current Opinion in Microbiology* 6, no. 5 (Oct. 2003): 452–56; H. Harbottle et al., "Genetics of Antimicrobial Resistance," *Animal Biotechnology* 17, no. 2 (2006): 111–24; L. D. Hogberg, A. Heddini, and O. Cars, "The Global Need for Effective Antibiotics: Challenges and Recent Advances," *Trends in Pharmacological Sciences* 31, no. 11 (2010): 509–15; A. O. Summers, "Genetic Linkage and Horizontal Gene Transfer, the Roots of the Antibiotic *Multi*-Resistance Problem," *Animal Biotechnology* 17, no. 2 (2006): 125–35.
64. NRDC, "Saving Antibiotics: What You Need to Know About Antibiotics Abuse on Farms," Feb. 7, 2014, http://www.nrdc.org/food/saving-antibiotics.asp.
65. Andrew Pollack, "Rising Threat of Infections Unfazed by Antibiotics," *New York Times*, Feb. 27, 2010.
66. Andrew E. Walters et al., "Multidrug-Resistant Staphylococcus aureus in

U.S. Meat and Poultry," *Clinical Infectious Diseases* 52, no. 10 (Feb. 2011): 1227–30.

67. Marissa Cevallos, "Meat Industry Fires Back at Drug-Resistant Staph Study," *Los Angeles Times,* Apr. 15, 2011.
68. Sabrina Tavernise, "Farm Use of Antibiotics Defies Scrutiny," *New York Times*, Sept. 3, 2012.
69. Sabrina Tavernise, "Antibiotics in Animals Tied To Risk of Human Infection," *New York Times*, Jan. 28, 2014.
70. L. Trasande et al., "Infant Antibiotic Exposures and Early-Life Body Mass," *International Journal of Obesity-Nature* 37, no. 1 (Jan. 2013): 16–23.
71. Pagan Kennedy, "The Fat Drug," *New York Times*, Mar. 9, 2014.
72. Julia Olmstead, "Bugs in the System: How the FDA Fails to Regulate Antibiotics in Ethanol Production," IATP, May 1, 2012. See also Tom Laskawy, "Do the Feds Care About Antibiotics in Animal Feed?" *Grist*, May 2, 2012; Tom Laskawy, "Three Reasons to Have a Cow over Antibiotics in Your Meat," *Grist*, Oct. 19, 2012; Maryn McKenna, "Antibiotics in Ethanol Grains: Glass Half-Empty or Half-Full? *Wired*, Apr. 10, 2012.
73. But it was clear even thirty years ago, to then FDA commissioner Donald Kennedy and other experts, that something should be done. Donald Kennedy, "Cows on Drugs," *New York Times* op-ed, Apr. 18, 2010.
74. Europa, "Ban on Antibiotics as Growth Promoters in Animal Feed Enters into Effect," press release, Dec. 22, 2005, http://europa.eu/rapid/press-release_IP-05-1687_en.htm.
75. Marilyn C. Roberts, "Using Antibiotics in Animal Feed Encourages Drug-Resistant Bacteria," *Seattle Times* op-ed, Apr. 21, 2013.
76. AVMA, "Extralabel Drug Use and AMDUCA: FAQ," https://www.avma.org/KB/Resources/FAQs/Pages/ELDU-and-AMDUCA-FAQs.aspx.
77. FDA, Animal & Veterinary, "Cephalosporin Order of Prohibition Questions and Answers," http://www.fda.gov/AnimalVeterinary/NewsEvents/CVMUpdates/ucm054434.htm.
78. *Natural Resources Defense Council Inc. v. U.S. Food and Drug Administration*, 2012 WL 983544 (S.D.N.Y. 2012), 872 F. Supp. 2d 318, 341–2, http://www.leagle.com/decision/In%20FDCO%2020120605A69.
79. Peter Lehner, "Ending FDA Paralysis on Antibiotics with Two Court Victories and a Push for Transparency," NRDC *Switchboard*, June 20, 2012, http://switchboard.nrdc.org/blogs/plehner/ending_fda_paralysis_on_antibi.html.
80. Lisa Heinzerling, "Undue Process at the FDA: Antibiotics, Animal Feed, and Agency Intransigence," *Vermont Law Review* 37, no. 4 (Summer 2013): 1007–31.

81. *Natural Resources Defense Council Inc. et al. v. United States Food and Drug Administration, et al.*, 11 Civ. 3562, June 1, 2012, United States District Court, Southern District of New York, http://law.justia.com/cases/federal/district-courts/new-york/nysdce/1:2011cv03562/379739/106; Gardiner Harris, "Steps Set for Livestock Antibiotic Ban," *New York Times*, Mar. 24, 2010.
82. FDA News & Events, "FDA Takes Significant Steps to Address Antimicrobial Resistance," Dec. 11, 2013, http://www.fda.gov/AnimalVeterinary/NewsEvents/CVMUpdates/ucm378166.htm.
83. Sabrina Tavernise, "F.D.A. Restricts Antibiotics Use for Livestock," *New York Times*, Dec. 12, 2013.
84. Columbia University School of International and Public Affairs, "Scientific Analysis of the Preservation of Antibiotics for Medical Treatment Act of 2011," The Earth Institute, Master of Public Administration in Environmental Science and Policy Workshop in Applied Earth Systems Management, Summer 2011.
85. Marks, *Cesspools of Shame*, 25. See also Sören Thiele-Bruhn, "Pharmaceutical Antibiotic Compounds in Soils—A Review," *Journal of Plant Nutrition and Soil Science* 166, no. 2 (Apr. 2003): 145–67; S. A. McEwen and P. J. Fedorka-Cray, "Antimicrobial Use and Resistance in Animals," *Clinical Infectious Diseases* 34, Supp. 3 (June 2002): S93–106.
86. D. A. Fatta et al., "Analytical Methods for Tracing Pharmaceutical Residues in Water and Wastewater," *Trends in Analytical Chemistry* 26, no. 6 (June 2007): 515–33.
87. McEwen and Fedorka-Cray, "Antimicrobial Use and Resistance in Animals."
88. Infectious Diseases Society of America, "Adverse Reactions to Antibiotics Send Thousands of Patients to the ER," *ScienceDaily*, press release, Aug. 13, 2008, http://www.sciencedaily.com/releases/2008/08/080812135515.htm.
89. M. Ellin Doyle, "Veterinary Drug Residues in Processed Meats—Potential Health Risk," Food Research Institute, University of Wisconsin–Madison, Mar. 2006, http://fri.wisc.edu/docs/pdf/FRIBrief_VetDrgRes.pdf.
90. G. M. Jones, "On-Farm Tests for Drug Residues in Milk," Virginia Cooperative Extension, May 2009, http://pubs.ext.vt.edu/404/404-401/404-401.html.
91. The National Milk Producers Federation says: "Dairy beef from market cows and bob veal make up a large number of the animals on the [FSIS] 'Residue Violator List,' though the numbers of positive animals represent a small percentage of the number of dairy animals processed every year," *Milk and Dairy Beef Drug Residue Prevention: Producer Manual of Best Management Practices*, Mar. 2012, 12. USDA Office of Inspector General, "FSIS National

Residue Program for Cattle," Audit Report 24601-08-K (Mar. 2012), 12, 26. Deborah Cera, team leader of the FDA's drug residue compliance team, says 7.7 percent of cattle slaughtered are adult dairy cows, but they account for 7.7 percent of drug residue violations.

92. USDA Office of Inspector General, "FSIS National Residue Program for Cattle," 13, 26, 30.
93. William Newman, "F.D.A. and Dairy Industry Spar over Testing of Milk," *New York Times*, Jan. 25, 2011.
94. Damon P. Miller, "National Conference on Interstate Milk Shipments (NCIMS) Update," TAPF Annual Meeting, June 2012.
95. Newman, "F.D.A. and Dairy Industry Spar over Testing of Milk."
96. The study launched in 2012 will compare 900 samples from previous violators with 900 randomly collected samples. An independent lab will look for the presence of over 30 antibiotics, not just the beta lactam (penicillin family) drugs previously targeted.
97. Frederick C. Michel and Zhongtang Yu, "Reduction of Antibiotic-Resistant Bacteria Present in Food Animal Manures by Composting and Anaerobic Digestion," Ohio Agricultural Research and Development Center, SEEDS grant, no. 52, Mar. 2010, http://oardc.osu.edu/seeds/seed52.pdf.
98. Michelle A. McConnell et al., "A Comparison of IgG and IgG1 Activity in an Early Milk Concentrate from Non-Immunized Cows and a Milk from Hyperimmunized Animals," *Food Research International* 34, nos. 2–3 (2001): 255–61; Gerco den Hartog et al., "Modulation of Human Immune Responses by Bovine Interleukin-10," *PLoS ONE* 6, no. 3 (2011); University of Otago (New Zealand), "McConnell Lab Research: Research Overview, http://micro.otago.ac.nz/research/research-labs/mcconnell-lab-research.
99. M. K. Hsieh et al., "Correlation Analysis of Heat Stability of Veterinary Antibiotics by Structural Degradation, Changes in Antimicrobial Activity and Genotoxicity," *Veterinarni Medicina* 56, no. 6 (2011): 274–85.
100. Matt Jorgensen and Pat Hoffman, "On-Farm Pasteurization of Milk for Calves," University of Wisconsin Dairy Update, Dairy Team Extension, http://www.uwex.edu/ces/heifermgmt/documents/pasteurization.pdf.
101. Consumer Reports, GreenerChoices.org, label search results for "antibiotic free," http://www.greenerchoices.org/eco-labels/label.cfm?LabelID=102.
102. Amanda D. Cuéllar and Michael E. Webber, "Cow Power: the Energy and Emissions Benefits of Converting Manure to Biogas," *Environmental Research Letters* 3, no. 3 (2008), http://iopscience.iop.org/1748-9326/3/3/034002/fulltext/.
103. Q. Wang et al., "Economic Feasibility of Converting Cow Manure to Electricity: A Case Study of the CVPS Cow Power Program in Vermont," *Journal of Dairy Science* 94, no. 10 (Oct. 2011): 4937–49.
104. Dan Morain, "Cow-Rich California's Biogas Record Stinks," *Press Dem-*

ocrat (Santa Rosa, CA), Sept. 25, 2011, http://www.pressdemocrat.com/article/20110925/wire/110929710.

105. Diane Greer, "Economies of Scale in Renewable Natural Gas," *BioCycle* 50, no. 5 (May 2009): 39.
106. Deb Slater, "Farm Power!" KVOS TV 12 (Bellingham, WA), Mar. 13, 2010.
107. Researchers at Washington State University estimate that if half of the state's quarter-million cows were connected to digesters, up to 110 million pounds of methane (crudely equivalent to 1,183,750 tons of carbon dioxide) could be captured. Craig Frear et al., "Bioenergy and Bioproducts Fact Sheet: Vander Haak Dairy Anaerobic Digester," Washington State University, http://whatcom.wsu.edu/carbonmasters/documents/AD_VanderHaak_factsheet_7_13_2006.pdf.
108. Dulcey Simpkins, "Anaerobic Digestion FAQs," Michigan Biomass Energy Program, Sept. 2005, http://www.michigan.gov/documents/anaerobic_digester_FAQs_2005_137431_7.pdf.
109. J. A. Resende et al., "Prevalence and Persistence of Potentially Pathogenic and Antibiotic Resistant Bacteria During Anaerobic Digestion Treatment of Cattle Manure," *Bioresource Technology* 153 (Feb. 2014): 284–91.

Five
GOT MILKED?

1. USDA Dairy Data, Overview of Dairy Products: Per Capita Consumption, United States (Annual).
2. Hayden Stewart, Diansheng Dong, and Andrea Carlson, "Why Are Americans Consuming Less Fluid Milk? A Look at Generational Differences in Intake Frequency," USDA Economic Research Service, Report Number 19 (May 2013): 22.
3. Marissa Fessenden, "Boys and Girls May Get Different Breast Milk," *Scientific American* 307, no. 6 (Nov. 2012). It supports the Trivers-Willard hypothesis.
4. Melissa A. Larson et al., "Sexual Dimorphism Among Bovine Embryos in Their Ability to Make the Transition to Expanded Blastocyst and in the Expression of the Signaling Molecule IFN-τ," *Proceedings of the National Academy of Sciences* 98, no. 17 (2003): 9677–82.
5. Jessica A. Barzilay, "Study Links Fetal Gender to Milk Production in Cows," *The Harvard Crimson*, Feb. 7, 2014.
6. Fessenden, "Boys and Girls May Get Different Breast Milk."
7. Walter C. Willett et al., letter to Carole Davis, USDA Dietary Guidelines Advisory Committee, July 15, 2010.
8. Dairy Management Inc. operates under several names: http://www.dairy

.org/. See also James McWilliams, "How Journalists Got the Cheese Lobbying Story Wrong," *The Atlantic*, Nov. 17, 2010.

9. Quanhe Yang et al., "Added Sugar Intake and Cardiovascular Diseases Mortality Among U.S. Adults, *Journal of the American Medical Association*, Feb. 3, 2014.
10. Sanjay Basu et al., "The Relationship of Sugar to Population-Level Diabetes Prevalence: An Econometric Analysis of Repeated Cross-Sectional Data, *PLoS ONE* 8, no. 2 (Feb. 27, 2013).
11. Dietary advice on saturated fats is confusing. The majority medical opinion is that they are to be avoided. But a new study involved 150 men and women who were divided into two groups: low-fat/high carbohydrates diets and high-fat/low carb diets (calorie intake and vegetable intake weren't limited). Those in the high-fat intake group were told to eat mostly protein and fat, and to favor foods with mostly unsaturated fats, but they were allowed to eat cheese, butter, and red meat. People in the high-fat group lost more weight and saw their HD ("good" cholesterol) rise and levels of triglycerides fall. The decline could have been due to their greater loss of weight (eight pounds more), the diet's short length (one year), and that participants got regular diet coaching and replaced one meal a day with a specified shake or bar. The high-fat/low-carb group had greater trouble sticking to their meaty diet. Additionally, the low-fat/high-carb group was asked to reduce their consumption of fat by only $\frac{1}{7}$, while the other group tried to reduce their carbohydrate consumption by $\frac{5}{6}$; this might help explain the differences in weight loss. Lydia A. Bazzano et al., "Effects of Low-Carbohydrate and Low-Fat Diets: A Randomized Trial," *Annals of Internal Medicine* 161, no. 5 (Sep. 2014): 309–18.

    Immediately after this study was released, a study with opposing findings hit the news. A meta-analysis of fifty other research papers, it found no weight difference after people were on a low-fat or low-carb diet for one year. About 7,300 volunteers were involved in the studies that comprised the meta-analysis. This and other large-scale studies conclude that the weight-loss diet you can stick with is the best diet for you. Bradley C. Johnston et al., "Comparison of Weight Loss Among Named Diet Programs in Overweight and Obese Adults: A Meta-analysis," *Journal of the American Medical Association* 312, no. 9 (Sep. 3, 2014): 923–33.
12. Harvard School of Public Health, The Nutrition Source, "Top Food Sources of Saturated Fat in the U.S.," http://www.hsph.harvard.edu/nutritionsource/top-food-sources-of-saturated-fat-in-the-us/.
13. International Dairy Foods Association, "Vanilla Top Ice Cream Flavor with Americans; Frozen Yogurt Regaining Fans," news release, June 21, 2012.

14. David S. Ludwig and Walter C. Willett, "Three Daily Servings of Reduced-Fat Milk: An Evidence-Based Recommendation?" *JAMA Pediatrics* 167, no. 9 (Sept. 2013): 788–89.
15. J. L. Maguire et al., "The Relationship Between Cow's Milk and Stores of Vitamin D and Iron in Early Childhood," *Pediatrics* 131, no. 1 (Jan. 2013): 144–51.
16. Kate Clancy, *Greener Pastures: How Grass-Fed Beef and Milk Contribute to Healthy Eating* (Cambridge, MA: Union of Concerned Scientists, 2006), 45 (steak), 49 (milk). See also Jana Kraft et al., "Extensive Analysis of Long-Chain Polyunsaturated Fatty Acids, CLA, *trans*-18:1 Isomers, and Plasmalogenic Lipids in Different Retail Beef Types," *Journal of Agriculture and Food Chemistry* 56, no. 12 (June 2008): 4775–82; S. K. Duckett et al., "Effects of Winter Stocker Growth Rate and Finishing System on: III. Tissue Proximate, Fatty Acid, Vitamin, and Cholesterol Content," *Journal of Animal Science* 87, no. 9 (Sept. 2009).
17. Liesbeth A. Smit et al., "Conjugated Linoleic Acid in Adipose Tissue and Risk of Myocardial Infarction," *American Journal of Clinical Nutrition* 92, no. 1 (July 2010): 34–40.
18. K. W. Lee et al., "Role of the Conjugated Linoleic Acid in the Prevention of Cancer," *Critical Reviews in Food Science and Nutrition* 45, no. 2 (2005): 135–44.
19. A. Bhattacharya et al., "Biological Effects of Conjugated Linoleic Acids in Health and Disease," *Journal of Nutritional Biochemistry* 17, no. 12 (Dec. 2006): 789–810.
20. M. A. Zulet et al., "Inflammation and Conjugated Linoleic Acid: Mechanisms of Action and Implications for Human Health," *Journal of Physiology and Biochemistry* 61, no. 3 (Sept. 2005): 483–94.
21. Sabine Tricon et al., "Opposing Effects of *cis*-9,*trans*-11 and *trans*-10,*cis*-12 Conjugated Linoleic Acid on Blood Lipids in Healthy Humans," *American Journal of Clinical Nutrition* 80, no. 3 (Sept. 2004): 614–20.
22. L. D. Whigham, A. C. Watras, and A. Schoeller, "Efficacy of Conjugated Linoleic Acid for Reducing Fat Mass: A Meta-Analysis in Humans," *American Journal of Clinical Nutrition* 85, no. 5 (May 2007): 1203–11. See also Magdalena Rosell, Niclas N. Håkansson, and Alicja Wolk, "Association Between Dairy Food Consumption and Weight Change over 9 Years in 19,352 Premenopausal Women," *American Journal of Clinical Nutrition* 84, no. 6 (Dec. 2006): 1481–88; and Andrea R. Josse et al., "Increased Consumption of Dairy Foods and Protein During Diet- and Exercise-Induced Weight Loss Promotes Fat Mass Loss and Lean Mass Gain in Overweight and Obese Premenopausal Women," *Journal of Nutrition*, July 20, 2011. This

study was co-funded by the Dairy Research Institute, the Dairy Farmers of Canada, and the Canadian Institutes of Health Research.

23. National Dairy Council, "Key Findings: National Medical Association Consensus Report," 2004, http://www.nationaldairycouncil.org/EDUCATIONMATERIALS/HEALTHPROFESSIONALSEDUCATIONKITS/Pages/KeyFindingsNMAConsensusReport.aspx, accessed July 19, 2013.
24. M. B. Zemel et al., "Calcium and Dairy Acceleration of Weight and Fat Loss During Energy Restriction in Obese Adults," *Obesity Research* 12, no. 4 (2004): 582–90.
25. David Schardt, "Milking the Data," *Nutrition Action Healthletter*, Sept. 2005. His results might apply only to people who are consuming too little calcium to begin with, and/or people who are overweight, or people not on a high-protein diet.
26. Melanie Warner, "Chug Milk, Shed Pounds? Not So Fast," *New York Times*, June 21, 2005.
27. Stephan Guyenet, "New Review Paper by Yours Truly: High-Fat Dairy, Obesity, Metabolic Health and Cardiovascular Disease," *Whole Health Source*, July 22, 2012, http://wholehealthsource.blogspot.com/2012/07/new-review-paper-by-yours-truly-high.html.
28. Frank A. Oski, *Don't Drink Your Milk* (Brushton, NY: TEACH Services, 2010).
29. Arne Høst, "Frequency of Cow's Milk Allergy in Childhood," *Annals of Allergy Asthma & Immunology* 89, no. 6, Supplement (Dec. 2002): 33–37.
30. A. Paddack et al., "Food Hypersensitivity and Otolaryngologic Conditions in Young Children," *Otolaryngology—Head and Neck Surgery* 147, no. 2 (Aug. 2012).
31. Høst, "Frequency of Cow's Milk Allergy in Childhood."
32. A. V. Schwartz et al., "International Variation in the Incidence of Hip Fractures: Cross-National Project on Osteoporosis for the World Health Organization Program for Research on Aging," *Osteoporosis International* 9, no. 3 (1999): 242–53; S. Maggi et al., "Incidence of Hip Fractures in the Elderly: A Cross-National Analysis," *Osteoporosis International* 1, no. 4 (Sept. 1991): 232–41.
33. U.S. Department of Health and Human Services, National Institutes of Health, Office of Dietary Supplements, "Dietary Supplement Fact Sheet: Calcium."
34. Connie M. Weaver and Karen L. Plawecki, "Dietary Calcium: Adequacy of a Vegetarian Diet," *American Journal of Clinical Nutrition* 59 (1994): 1238S-41S. Greens high in oxalates include spinach, beet greens, and Swiss chard. Oxalates make it more difficult for your body to absorb calcium. Academy of

Nutrition and Dietetics, "Eat Right—How Can I Get Enough Calcium in My Diet If I Follow a Vegan Lifestyle?" http://www.eatright.org/Public/content.aspx?id=10658.

35. H. A. Bischoff-Ferrari et al., "Calcium Intake and Hip Fracture Risk in Men and Women: A Meta-Analysis of Prospective Cohort Studies and Randomized Controlled Trials," *American Journal of Clinical Nutrition* 86, no. 6 (Dec. 2007): 1780–90; J. Porthouse et al., "Randomized Controlled Trial of Calcium and Supplementation with Cholecalciferol (Vitamin D3) for Prevention of Fractures in Primary Care," *British Medical Journal* 330, no. 7498 (2005): 1003. Porthouse et al. concluded, "We found no evidence that calcium and vitamin D supplementation reduces the risk of clinical fractures in women with one or more risk factors for hip fracture."

    Diane Feskanich, Walter C. Willett, and Graham A. Colditz, "Calcium, Vitamin D, Milk Consumption, and Hip Fractures: a Prospective Study among Postmenopausal Women," *American Journal of Clinical Nutrition* 77, no. 2 (Feb. 2003): 504–11. This eighteen-year prospective analysis of over 72,000 postmenopausal women found that adequate vitamin D intake lowered the risk of hip fractures, but that neither milk nor a high-calcium diet did any good.

36. H. A. Bischoff-Ferrari et al., "Milk Intake and Risk of Hip Fracture in Men and Women: A Meta-Analysis of Prospective Cohort Studies," *Journal of Bone and Mineral Research* 26, no. 4 (Apr. 2011): 833–39.

37. Kuanrong Li et al., "Associations of Dietary Calcium Intake and Calcium Supplementation with Myocardial Infarction and Stroke Risk and Overall Cardiovascular Mortality in the Heidelberg Cohort of the European Prospective Investigation into Cancer and Nutrition Study (EPIC-Heidelberg)," *Heart* 98, no. 920 (May 2012): 920–25. Vitamin D acts as a hormone, and its interrelationship with calcium is complicated, with many feedback loops. Much more research in needed to address many uncertainties. An excellent (1,132-page) source for calcium/vitamin D geeks is A. Catherine Ross et al., *Dietary Reference Intakes for Calcium and Vitamin D*, National Academies Press, 2011. It may be downloaded for free at http://books.nap.edu/catalog.php?record_id=13050&utm_expid=4418042-5.krRTDpXJQISoXLpdo-1Ynw.0&utm_referrer=http%3A%2F%2Fbooks.nap.edu%2Fopenbook.php%3Frecord_id%3D13050%26page%3D516.

38. U.S. Preventive Services Task Force, "Vitamin D and Calcium Supplementation to Prevent Fractures," Feb. 2013, http://www.uspreventiveservicestaskforce.org/uspstf/uspsvitd.htm.

39. Tom Laskawy, "Oh Rot, the White House Just Gutted the New Food Safety Rules," *Grist*, Apr. 4, 2012.

40. Cornell University Food Science Department, "Position on Raw Milk Sales and Consumption," http://www.marlerclark.com/images/uploads/about_ecoli/RAW-MILK-MQIP-Position-Statement-01-09.pdf.
41. CDC, "Raw Milk Questions and Answers," updated Mar. 7, 2014, http://www.cdc.gov/foodsafety/rawmilk/raw-milk-questions-and-answers.html.
42. L. E. Macdonald et al., "A Systematic Review and Meta-Analysis of the Effects of Pasteurization on Milk Vitamins, and Evidence for Raw Milk Consumption and Other Health-Related Outcomes," *Journal of Food Protection* 74, no. 11 (Nov. 2011): 1814–32. The authors caution that the methodology of many of the studies they looked at was poor. See also Marler Blog, "Raw Milk Pros: Review of the Peer-Reviewed Literature," June 6, 2008, http://www.marlerblog.com/lawyer-oped/raw-milk-pros-review-of-the-peer-reviewed-literature/#.U4svL_ldXa4.
43. Quyen Vu and Sarah Mummah, "Effect of Raw Milk on Lactose Intolerance Symptoms; a Randomized Controlled Trial," link to poster at http://www.marlerblog.com/case-news/stanford-bites-raw-milk-in-the-udder/#.UqTbuxyOKqU.
44. Cornell University Food Science Department, "Position Statement on Raw Milk Sales and Consumption," http://www.milkfacts.info/Current%20Events/Position%20Statement%20%20Raw%20Milk.pdf.
45. *Code of Federal Regulations*, Cheeses and Related Cheese Products, title 21, part 133, http://www.accessdata.fda.gov/scripts/cdrh/cfdocs/cfcfr/CFRSearch.cfm?CFRPart=133.
46. Bill Marler, "A Look at the Safety—Or Not—of Raw Milk Cheese and the "60 Day Rule," Sept. 19, 2013, http://www.marlerblog.com/lawyer-oped/a-look-at-the-safety-or-not-of-raw-milk-cheese-and-the-60-day-rule/#.U4su4PldXa4.
47. William Neuman, "Raw Milk Cheesemakers Fret Over Possible New Rules," *New York Times*, Feb. 4, 2011.
48. FDA News and Events, http://www.fda.gov/NewsEvents/Newsroom/PressAnnouncements/2010/ucm232748.htm; William Neuman, "Raw Milk Cheesemakers Fret over Possible New Rules," *New York Times*, Feb. 4, 2011.
49. David Gumpert, "FDA's Crackdown on Raw-Milk Cheese Based on Flawed Data Analysis," *Grist*, Feb. 18, 2011.
50. John Hermon-Taylor, "*Mycobacterium avium* Subspecies *paratuberculosis*, Crohn's Disease and the Doomsday Scenario," *Gut Pathogens* 1, no. 15 (July 2009). Hermon-Taylor owns patents to a vaccine against MAP. Entire article: http://www.ncbi.nlm.nih.gov/pmc/articles/PMC2718892/?tool=pubmed.
51. D. L. Clark et al., "Detection of *Mycobacterium avium* Subspecies *paratuberculosis* Genetic Components in Retail Cheese Curds Purchased in Wisconsin

and Minnesota by PCR," *Molecular and Cellular Probes* 20, no. 3–4 (June–Aug. 2006): 197–202; J. L. Ellingson et al., "Detection of Viable *Mycobacterium avium* subsp. *paratuberculosis* in Retail Pasteurized Whole Milk by Two Culture Methods and PCR," *Journal of Food Protection* 68 (2005): 966–72.

52. USDA/APHIS, "Johne's Disease on U.S. Dairies, 1991–2007, http://www.aphis.usda.gov/animal_health/nahms/dairy/downloads/dairy07/Dairy07_is_Johnes.pdf."
53. Ellen S. Pierce, "Possible Transmission of *Mycobacterium avium* Subspecies *paratuberculosis* through Potable Water: Lessons from an Urban Cluster of Crohn's Disease," *Gut Pathogens* 1, no. 17 (2009), http://www.gutpathogens.com/content/1/1/17.
54. J. D. Sanderson et al., "*Mycobacterium paratuberculosis* DNA in Crohn's Disease Tissue," *Gut* 33 (1992): 890–96, http://www.ncbi.nlm.nih.gov/pmc/articles/PMC1379400/pdf/gut00574-0040.pdf.
55. Pierce, "Possible Transmission of *Mycobacterium avium* Subspecies *paratuberculosis* through Potable Water: Lessons from an Urban Cluster of Crohn's Disease.
56. Michael Collins and Elizabeth Manning, "Zoonotic Potential at a Glance," Johne's Information Center, University of Wisconsin, School of Veterinary Medicine, updated Mar. 2010, http://www.johnes.org/zoonotic/index.html.
57. Lawrence M. Wein and Yifan Liu, "Analyzing a Bioterror Attack on the Food Supply: The Case of Botulinum Toxin in Milk," *Proceedings of the National Academy of Sciences* 102, no. 28 (July 2005).
58. Lawrence M. Wein, "Got Toxic Milk?" *New York Times* op-ed, May 30, 2005, http://www.nytimes.com/2005/05/30/opinion/30wein.html?pagewanted=all&_r=0.
59. A twenty-eight-page manual called "Preparation of Botulism Toxin" is available on many jihadist websites, and the toxin itself can be obtained from overseas black-market laboratories.
60. Oliver G. Weingart et al., "The Case of Botulinum Toxin in Milk: Experimental Data," *Applied and Environmental Microbiology* 76, no. 10 (May 2010): 3293–3300.
61. Bruce Alberts, "Modeling Attacks on the Food Supply," *Proceedings of the National Academy of Sciences* 102, no. 28 (June 2005): 9737–38.
62. Corydon Ireland, "Hormones in Milk Can Be Dangerous," Harvard News Office, http://www.news.harvard.edu/gazette/2006/12.07/11-dairy.html; D. Ganmaa and A. Sato, "The Possible Role of Female Sex Hormones in Milk from Pregnant Cows in the Development of Breast, Ovarian and Corpus Uteri Cancers," *Medical Hypotheses* 65, no. 6 (Aug. 2005): 1028–37.
63. Edmund Sanders, "Israel Sperm Banks Find Quality Is Plummeting," *Los*

Angeles Times, Aug. 15, 2012, http://articles.latimes.com/2012/aug/15/world/la-fg-israel-sperm-20120816/2.

64. D. Ganmaa and A. Sato, "Consumption of Cow's Milk and Possible Risk of Breast Cancer," *Breast Care* 5, no. 1 (Mar. 2010): 44–46. Several studies have found a strong correlation between mortality from prostate cancer and heavy milk consumption. A report published by the Physicians Committee for Responsible Medicine concluded: "Evidence from international, case-control, and cohort studies suggests that men who avoid dairy products are at lower risk for prostate cancer incidence and mortality, compared to others," http://gna.squarespace.com/home/gary-null-effects-of-dairy.html. In women, milk and cheese consumption has been associated with ovarian and breast cancer, but not all researchers agree.
65. Harvard School of Public Health, The Nutrition Source, "Calcium and Milk: What's Best for Your Bones and Health?" http://www.hsph.harvard.edu/nutritionsource/calcium-full-story/.
66. FDA, "Report on the Food and Drug Administration's Review of the Safety of Recombinant Bovine Somatotropin," Apr. 23, 2009, http://www.fda.gov/AnimalVeterinary/SafetyHealth/ProductSafetyInformation/ucm130321.htm.
67. American Cancer Society, "Recombinant Bovine Growth Hormone," http://www.cancer.org/cancer/cancercauses/othercarcinogens/athome/recombinant-bovine-growth-hormone.
68. Dairy Council of California, "Dairy Facts/Types of Milk/rBST Free Milk," http://www.healthyeating.org/Milk-Dairy/Dairy-Facts/Types-of-Milk.aspx?Referer=dairycouncilofca, accessed July 12, 2013; M. H. Le Breton et al., "Detection of Recombinant Bovine Somatotropin in Milk and Effect of Industrial Processes on Its Stability," *Analytica Chimica Acta* 672, nos. 1–2 (July 2010): 45–49. Pasteurization destroys up to 95 percent of rBST/BST in milk.
69. Tom Laskawy, "Court Rules rBGH-Free Milk Is Better Than the Kind Produced with Artificial Hormones. Now What?" *Grist*, Oct. 6, 2010.
70. FDA, "Report on the Food and Drug Administration's Review of the Safety of Recombinant Bovine Somatotropin," http://www.fda.gov/animalveterinary/safetyhealth/productsafetyinformation/ucm130321.htm.
71. For example, the commissioner in charge of writing the rBST labeling guidelines, Michael Taylor, had previously worked for Monsanto for many years, http://www.sustainabletable.org/797/rbgh#gsc.tab=0.
72. A. S. Wiley, "Cow Milk Consumption, Insulin-Like Growth Factor-1, and Human Biology: A Life History Approach," *American Journal of Human Biology* 24, no. 2 (Mar.–Apr. 2012): 130–38. "IGF-I is a candidate bioactive

molecule linking milk consumption to more rapid growth and development, although the mechanism by which it may exert such effects is unknown"; "Routine milk consumption is an evolutionarily novel dietary behavior that has the potential to alter human life history parameters, especially vis-à-vis linear growth, which in turn may have negative long-term biological consequences."

73. Francesca L. Crowe et al., "The Association Between Diet and Serum Concentrations of IGF-1, IGFBP, IGFBP-2, and IGFBP-3 in the European Prospective Investigation into Cancer and Nutrition," *Cancer Epidemiology, Biomarkers & Prevention* 18, no. 5 (May 2009): 1333–40. "The results from this large cross-sectional analysis show that either the intake of dairy protein or calcium is an important dietary determinant of IGF-I and IGFBP-2 concentrations; however, we suggest that it is more likely to be protein from dairy products."
74. Health Canada, "Report of the Canadian Veterinary Medical Association Expert Panel on rBST," Nov. 1998, http://www.hc-sc.gc.ca/dhp-mps/vet/issues-enjeux/rbst-stbr/rep_cvma-rap_acdv_tc-tm-eng.php.
75. Shaoni Bhattacharya, "Study Reveals Chemical Cocktail in Every Person," *NewScientist*, Nov. 25, 2003.
76. Laura N. Vandenberg et al., "Hormones and Endocrine-Disrupting Chemicals: Low-Dose Effects and Nonmonotonic Dose Responses," *Endocrine Reviews* 33, no. 3 (June 2012): 378–455.
77. A. Schecter et al., "Intake of Dioxins and Related Compounds from Food in the U.S. Population," *Journal of Toxicology and Environmental Health, Part A* 63, no. 1 (May 2011): 1–18.
78. Center for Food Safety, "Genetically Modified (GM) Crops and Pesticide Use," May 2008, Appendix 2—"Usage of Leading Herbicides Other than Glyphosate on Corn and Soy in the U.S.: 2002 to 2006."
79. Environmental Working Group, Reports & Consumer Guides, "CDC Scientists Find Rocket Fuel Chemical in Infant Formula," Apr. 2, 2009, http://www.ewg.org/research/cdc-scientists-find-rocket-fuel-chemical-infant-formula.
80. Chris E. Talsness et al., "*In Utero* and Lactational Exposures to Low Doses of Polybrominated Diphenyl Ether-47 Alter the Reproductive System and Thyroid Gland of Female Rat Offspring," *Environmental Health Perspectives* 116, no. 3 (Mar. 2008): 308–14.
81. World Health Organization, "Dioxins and their Effects on Human Health," Fact Sheet No. 225, May 2010, http://www.who.int/mediacentre/factsheets/fs225/en/.
82. M. D. Anway et al., "Epigenetic Transgenerational Actions of Endocrine Disruptors," *Science* 3, no. 308 (June 2005): 1466–69; M. D. Amway et al.,

"Transgenerational Effect of the Endocrine Disruptor Vinclozolin on Male Spermatogenesis," *Journal of Andrology* 27, no. 6 (Nov.–Dec. 2006): 868–79; David Crews et al., "Epigenetic Transgenerational Inheritance of Altered Stress Responses," *Proceedings of the National Academy of Sciences* 109, no. 23 (June 2012). See also Theo Colburn, Dianne Dumanoski, and John Peterson Myers, *Our Stolen Future* (New York: Dutton/Penguin Books, 1996); Mark Schapiro, *Exposed: The Toxic Chemistry of Everyday Products and What's at Stake for American Power* (White River Junction, VT: Chelsea Green Publishing, 2007).

83. J. R. Roy, S. Chakraborty, and T. R. Chakraborty, "Estrogen-Like Endocrine Disrupting Chemicals Affecting Puberty in Humans—A Review," *Medical Science Monitor* 15, no. 6 (June 2009): RA137–45; Sandra Steingraber, *The Falling Age of Puberty in U.S. Girls: What We Know, What We Need to Know* (San Francisco: Breast Cancer Fund, 2007); Shari Roan, "Some Girls' Puberty Age Still Falling, Study Suggests," *Los Angeles Times*, Aug. 8, 2010. Increased rates of obesity may also play a role in these shortened childhoods.
84. Frank M. Biro et al., "Pubertal Assessment Method and Baseline Characteristics in a Mixed Longitudinal Study of Girls," *Pediatrics* 126, no. 3 (Sept. 2010), http://pediatrics.aappublications.org/content/126/3/e583.full.
85. Ryan Pandya, "Milk Without the Moo," *New Scientist*, June 28, 2014, http://www.newscientist.com/article/mg22229750.400-dont-have-a-cow-making-milk-without-the-moo.html.
86. Kurt Timmermeister, *Growing a Farmer: How I Learned to Live Off the Land* (New York, W. W. Norton, 2011).

Six
DON'T HAVE A COWBURGER

1. Jan Havlicek and Pavlina Lenochova, "The Effect of Meat Consumption on Body Odor Attractiveness," *Chemical Senses* 31, no. 8 (July 2006): 747–52.
2. Betty Fussell, *Raising Steaks: The Life and Times of American Beef* (New York: Mariner/Houghton Mifflin Harcourt, 2009), 5.
3. Leslie C. Aiello and Peter Wheeler, "The Expensive-Tissue Hypothesis: The Brain and the Digestive System in Human and Primate Evolution," *Current Anthropology* 36, no. 2 (Apr. 1995): 199–221.
4. The daily recommended amount for iron is 18 mg. U.S. Department of Health and Human Services, National Institutes of Health, Office of Dietary Supplements, "Dietary Supplement Fact Sheet: Iron." Three ounces of beef

liver provides 29 percent; 3 ounces of lean chuck or blade roast provides 17 percent; 3 ounces of 85 percent lean ground beef provides 12 percent.

5. Morgan E. Levine et al., "Low Protein Intake Is Associated with a Major Reduction in IGF-1, Cancer, and Overall Mortality in the 65 and Younger but Not Older Population," *Cell Metabolism* 19, no. 3 (Mar. 2014): 407–17, http://www.cell.com/cell-metabolism/pdf/S1550-4131(14)00062-X.pdf.
6. Suzanne Wu, "Meat and Cheese May Be as Bad for You as Smoking," *ScienceDaily*, Mar. 4, 2014, http://www.sciencedaily.com/releases/2014/03/140304125639.htm; Levine et al., "Low Protein Intake."
7. *HealthDay*, "High-Protein Diets in Middle Age Might Shorten Life Span," Mar. 4, 2014, http://consumer.healthday.com/vitamins-and-nutrition-information-27/food-and-nutrition-news-316/high-protein-diets-in-middle-age-might-shorten-life-span-685436.html.
8. Burger King menu, http://www.bk.com/en/us/menu-nutrition/lunch-and-dinner-menu-202/fire-grilled-burgers-and-sandwiches-220/index.html, accessed Mar. 21, 2014. A Harvard study found that adults, adolescents, and children all dramatically underestimate the calorie content of their fast food choices. Jason P. Block et al., "Consumers' Estimation of Calorie Content at Fast Food Restaurants: Cross Sectional Observational Study," *BMJ* 346:f2907 (May 23, 2013), http://www.bmj.com/content/346/bmj.f2907.
9. Christopher Davis and Biing-Hwan Lin, "Factors Affecting U.S. Beef Consumption," USDA Economic Research Service Outlook Report LDP-M-135-02, Oct. 2005, 7.
10. Ibid., 1, 8, 10, 15, 16, 19.
11. CDC, "Protein," http://www.cdc.gov/nutrition/everyone/basics/protein.html.
12. Darius Lakdawalla and Tomas Philipson, "The Growth of Obesity and Technological Change: A Theoretical and Empirical Examination," National Bureau of Economic Research, working paper no. 8946 (May 2002).
13. CDC, "The Health Effects of Overweight and Obesity," http://www.cdc.gov/healthyweight/effects/index.html.
14. Ryan K. Masters et al., "The Impact of Obesity on U.S. Mortality Levels: The Importance of Age and Cohort Factors in Population Estimates," *American Journal of Public Health*, Oct. 2013.
15. The Continuous Update Project (CUP), a joint effort by the World Cancer Research Fund and the American Institute for Cancer Research, concluded that there is convincing scientific evidence that red meat and processed meat increase the risk of colon cancer. See also Doris S. M. Chan et al., "Red and Processed Meat and Colorectal Cancer Incidence: Meta-Analysis of Prospective Studies," *PLoS ONE* 6, no. 6 (June 2011).

16. An Pan et al., "Red Meat Consumption and Mortality: Results from 2 Prospective Cohort Studies," *Archives of Internal Medicine* 172, no. 7 (Apr. 9, 2012): 555–63.
17. Ibid. Most participants were non-Hispanic white health professionals, so the results might not apply to other populations. Other limitations are that the study was based on what people *reported* eating, and increased smoking, drinking alcohol to excess, and exercise avoidance may have made meat consumption look unhealthier, even though the researchers tried to control for smoking and obesity. See also Nicholas Bakalar, "Risks: More Red Meat, More Mortality," *New York Times*, Mar. 12, 2012; *ScienceDaily*, "Red Meat Consumption Linked to Increased Risk of Total, Cardiovascular, and Cancer Mortality," Mar. 12, 2012. A Harvard meta-study reviewed twenty studies involving 1.2 million meat eaters and concluded, "The overall findings suggest that neither unprocessed red nor processed meat consumption is beneficial for cardiometabolic health. . . .": R. Micha, G. Michas, and D. Mozaffarian, "Unprocessed Red and Processed Meats and Risk of Coronary Artery Disease and Type 2 Diabetes—an Updated Review of the Evidence," *Current Atherosclerosis Reports* 14, no. 6 (Dec. 2012): 515–24.
18. P. W. Siri-Tarino et al., "Saturated Fatty Acids and Risk of Coronary Heart Disease: Modulation by Replacement Nutrients," *Current Atherosclerosis Reports* 12, no. 6 (Nov. 2012): 384–90.
19. Teresa Norat et al., "The Associations Between Food, Nutrition and Physical Activity and the Risk of Colorectal Cancer," WCRF/AICR Systematic Literature Review, Continuous Update Project Report (Oct. 2010): 124.
20. S. Y. Foo et al., "Vascular Effects of a Low-Carbohydrate High-Protein Diet," *Proceedings of the National Academy of Sciences* 106, no. 36 (Sept. 2009): 15418–23; Renata B. Kostogrys et al., "Low Carbohydrate, High Protein Diet Promotes Atherosclerosis in Apolipoprotein E/Low-Density Lipoprotein Receptor Double Knockout Mice (apoE/LDLR $(^{-/-})$)," *Atherosclerosis* 223, no. 2 (Aug. 2012): 327–31.
21. Ferris Jabr, "Meat of the Matter: Are Our Modern Methods of Preserving and Cooking Meat Healthy? Why Steaks Could Be In, but Hot Dogs Still Out," *Scientific American*, Nov. 2012. There is some evidence that high-salt diets might *not* make most white folks likely to die of heart disease or get high blood pressure. Genevra Pittman, "Eating Less Salt Doesn't Cut Heart Risks: Study," Reuters, May 3, 2011.
22. B. J. Bennett et al., "Trimethylamine-N-oxide, a Metabolite Associated with Atherosclerosis, Exhibits Complex Genetic and Dietary Regulation," *Cell Metabolism* 17, no. 1 (Jan. 2013): 49–60; Gina Kolata, "Culprit in Heart Disease Goes Beyond Meat's Fat," *New York Times,* Apr. 8, 2013; W. H. Wilson

Tang et al., "Intestinal Microbial Metabolism of Phosphatidylcholine and Cardiovascular Risk," *New England Journal of Medicine* 368, no. 17 (Apr. 2013): 1575–84; Brie Zeltner, "Compound in Red Meat, Energy Drinks Linked to Heart Disease in Cleveland Clinic Research, *Plain Dealer*, Apr. 7, 2013.

23. Santica M. Marcovina et al., "Translating the Basic Knowledge of Mitochondrial Functions to Metabolic Therapy: Role of L-Carnitine," *Translational Research* 161, no. 2 (Feb. 2013): 73–84.
24. USDA Center for Nutrition Policy and Promotion, "Nutrient Content of the U.S. Food Supply, 1909–2004," Home Economics Research Report No. 57, Feb. 2007.
25. Sanjay Gupta for CBS, *60 Minutes*, "Is Sugar Toxic?" Apr. 1, 2012; Gary Taubes, "Is Sugar Toxic?" *New York Times Magazine*, Apr. 13, 2011.
26. Clancy, *Greener Pastures*, 20; Jana Kraft et al., "Extensive Analysis of Long-Chain Polyunsaturated Fatty Acids, CLA, *trans*-18:1 Isomers, and Plasmalogenic Lipids in Different Retail Beef Types," *Journal of Agriculture and Food Chemistry* 56, no. 12 (June 2008): 4775–82; S. K. Duckett et al., "Effects of Winter Stocker Growth Rate and Finishing System on: III. Tissue Proximate, Fatty Acid, Vitamin, and Cholesterol Content," *Journal of Animal Science* 87, no. 9 (Sept. 2009): 2961–70. J. M. Leheska et al., "Effects of Conventional and Grass Feeding Systems on the Nutrient Composition of Beef," *Journal of Animal Science* 86, no. 12 (Dec. 2008): 3575–85.
27. Olivia I. Okereke et al., "Dietary Fat Types and 4-Year Cognitive Change in Community-Dwelling Older Women," *Annals of Neurology* 72, no. 1 (July 2012): 124–34. See also Marie-Noël Vercambre, Francine Grodstein, and Jae Hee Kang, "Dietary Fat Intake in Relation to Cognitive Change in High-Risk Women with Cardiovascular Disease or Vascular Factors," *European Journal of Clinical Nutrition* 72, no. 1 (July 2012): 124–34.
28. Brian Montopoli, "Poll: Most Not Fully Confident in Food Safety," CBS News, Jan. 9, 2010.
29. Fred Pearce, "Earth's Nine Life-Support Systems: Chemical Pollution," *NewScientist,* Feb. 24, 2010.
30. For an overview, see David Ewing Duncan, "Chemicals Within Us," *National Geographic*, May 2012.
31. Schapiro, *Exposed: The Toxic Chemistry of Everyday Products and What's at Stake for American Power*, 132.
32. *New York Times* editorial staff, "A Toothless Law on Toxic Chemicals," *New York Times*, Apr. 19, 2013.
33. The Federal Insecticide, Fungicide, and Rodenticide Act (FIFRA), also administered by the EPA, ordered a onetime reevaluation of older pesticides to ensure they met current standards. The process (called reregistration) was

completed in 2008, but implementation continues. FIFRA denied registration unless a chemical "will not generally cause unreasonable adverse effects on the environment." This included harm to humans who eat crops or meat treated with an insecticide, fungicide, or rodenticide. The law shifted the burden of proving a pesticide safe to the manufacturer. Under another EPA program, the active ingredients in all pesticides are to be reevaluated every fifteen years ("registration review"). But the EPA is instructed to balance any harm caused against the economic benefits of continued usage. EPA, Reevaluation: Review of Registered Pesticides, Feb. 2013, http://www.epa.gov/oppsrrd1/reevaluation/environmental-hazard-statment.html.

34. CDC, *Fourth National Report on Human Exposure to Environmental Chemicals* (2009); Updated Tables (Mar. 2013), http://www.cdc.gov/exposurereport/.
35. S. H. Swan et al., "Semen Quality of Fertile U.S. Males in Relation to Their Mothers' Beef Consumption During Pregnancy," *Human Reproduction* 22, no. 6 (June 2007): 1497–1502.
36. S. H. Swan, "Semen Quality in Fertile U.S. Men in Relation to Geographical Area and Pesticide Exposure," *International Journal of Andrology* 29, no. 1 (Feb. 2006): 62–68.
37. USDA Office of Inspector General, Audit Report 24601-08-KC, Mar. 2010.
38. Ron Nixon, "Shipping Continued After Computer Inspection System Failed at Meat Plants," *New York Times*, Aug. 18, 2013.
39. Janet Raloff, "Hormones: Here's the Beef. Environmental Concerns Reemerge over Steroids Given to Livestock," *Science News* 161, no. 1 (Jan. 2002).
40. Montgomery, *A Cow's Life*, 147.
41. W. J. Platter et al., "Effects of Repetitive Use of Hormonal Implants on Beef Carcass Quality, Tenderness, and Consumer Ratings of Beef Palatability," *Journal of Animal Science* 81, no. 4 (Apr. 2003): 984–96.
42. FDA, "Steroid Hormone Implants Used for Growth in Food-Producing Animals," updated Feb. 8, 2011, http://www.fda.gov/animalveterinary/safetyhealth/productsafetyinformation/ucm055436.htm.
43. Opinion of the Scientific Committee on Veterinary Measures Relating to Public Health, "Assessment of the Potential Risks to Human Health from Hormone Residues in Bovine Meat and Meat Products," European Commission, Apr. 1999. The panel reaffirmed its findings in 2000 and 2002. Associated Press, "EU Scientists Confirm Health Risks of Growth Hormones in Meat," Apr. 23, 2002. Available at Organic Consumers Association website, http://www.purefood.org/toxic/hormone042302.cfm.
44. Geoff Winestock, "Column One: Beef Battle with U.S. Began in a Rare

Event—Italian Boys Grew Breasts; Ministers Banned Hormones—But Scientific Evidence Was Always Lacking," *Wall Street Journal Europe*, Mar. 2, 2000.

45. S. Scaglioni et al., "Breast Enlargement at an Italian School," *The Lancet* 1, no. 8063 (1978): 551–52.
46. University of California at Berkeley, *The New Wellness Encyclopedia* (1995), 202.
47. Laurence Peter, "Trade Deal Eases EU-U.S. Beef War over Hormones, BBC News Europe, Mar. 14, 2012.
48. Lindsay Abrams, "New Concern over Additives in Cattle Feed," *Salon*, Aug. 13, 2012.
49. Lisa Baertlein and P. J. Huffstutter, "Insight: Some U.S. Feedlots Rue Loss of 'Vitamin K' Zilmax," Reuters, Aug. 18, 2013.
50. University of Nebraska Extension, "Beta-Agonists: What Are They and Should I Be Concerned?" *Bovine Veterinarian*, updated Oct. 2013, http://newsroom.unl.edu/announce/beef/2563/14863.
51. Kelsey Tee, "The Drugs in Our Beef: Bigger Cows, More Worries," *Wall Street Journal*, Aug. 8, 2013, http://blogs.wsj.com/corporate-intelligence/2013/08/08/the-drugs-in-our-beef-bigger-cows-more-worries/.
52. H. A. Kuiper et al., "Illegal Use of ß-Adrenergic Agonists: European Community," *Journal of Animal Science* 76, no. 195 (Jan. 1998): 195–207; G. A. Mitchell and G. Dunnavan, "Illegal Use of Beta-Adrenergic Agonists in the United States," *Journal of Animal Science* 76, no. 1 (Jan. 1998): 208–11.
53. Tom Polansek and P. J. Huffstutter, "RPT—Halt in Zilmax Sales Fuels Demand for Rival Cattle Feed Product," Reuters, Aug. 26, 2013.
54. John Maday, "Merck Updates Veterinarians on Zilmax," *Drovers CattleNetwork*, Dec. 5, 2013.
55. FDA, "Freedom of Information Summary, NADA 141–223, Optaflexx 45," Dec. 11, 2009.
56. USDA, "Bison from Farm to Table," http://www.fsis.usda.gov/wps/portal/fsis/topics/food-safety-education/get-answers/food-safety-fact-sheets/meat-preparation/focus-on-bison/CT_Index, accessed June 1, 2014.
57. Tallgrass prairie in the United States existed just east of the Great Plains and was blessed with good loess soil and moderate rainfall. The short-grass prairie typical of the Great Plains had less rain and worse soil. Its typical grasses were buffalo-grass and blue grama. Bison inhabited both types of prairie but are most closely associated with the short-grass ecosystem.
58. Valerius Geist, *Buffalo Nation* (Minneapolis: Voyageur Press, 1998), 91, describes Sheridan as saying this to a joint session of the Texas Senate and House in 1875 when they were considering a bill to protect the few remaining bison. The bill did not pass.

59. Nate Schweber, "As Bison Return to Prairie, Some Rejoice, Others Worry," *New York Times*, Apr. 27, 2012.
60. Jim Robbins, "On the Montana Range, Efforts to Restore Bison Meet Resistance," *New York Times*, Apr. 2, 2013.
61. USDA, "Bison from Farm to Table."
62. James Derr, "American Bison: The Ultimate Genetic Survivor," slide show for Texas A&M University students, Oct. 24, 2006.
63. American Beefalo Association, "Beefalo Nutritional Facts," http://americanbeefalo.org/beefalo-nutritional-facts/.
64. When species interbreed, the mitochondrial and nuclear genomes their offspring end up with might be somewhat incompatible, resulting in less healthy animals. If the mtDNA mutations aren't too severe, the weaker animals can still breed, passing along the inferior mtDNA.
65. George Wuerthner, "Just a Domestic Bison? Cattle Are No Substitute for Buffalo," in *Welfare Ranching: The Subsidized Destruction of the American West*, edited by George Wuerthner and Mollie Matteson (Washington, D.C.: Island Press, 2002): 295–97.
66. Bryce Andrews, *Badluck Way: A Year on the Ragged Edge of the West* (New York: Altria Publishing Group, 2014).
67. The act was not enacted before Congress adjourned. Peter Applebome, "A Bison So Rare It's Sacred," *New York Times*, July 12, 2012.
68. "Hundreds Celebrate Rare White Bison at Conn. Farm," AP/*Seattle Post-Intelligencer*, July 28, 2012.
69. Susan E. Wolver et al., "A Peculiar Cause of Anaphylaxis: No More Steak? The Journey to Discovery of a Newly Recognized Allergy to Galactose-alpha-1,3-Galactose Found in Mammalian Meat," *Journal of General Internal Medicine* 28, no. 2 (Feb. 2013): 322–25; Olive Smith, "Ticks Causing Mysterious Meat Allergy," CNN, June 20, 2012.
70. Mary Esselman, "Profile: Author John Grisham's Allergy Mystery," *Allergic Living*, April 2012, http://allergicliving.com/2012/04/10/profile-author-john-grishams-allergy-mystery/.
71. Tom Jensen, "Food Issues Polarizing America," Public Policy Polling, Feb. 26, 2013. Polls of this sort report inconsistent results.
72. Ben Adler, "Al Gore Is a Vegan Now—And We Think We Know Why," *Grist*, Nov. 27, 2013.
73. Henry Fountain, "Building a $325,000 Burger," *New York Times*, May 14, 2012; Ian Sample, "£200,000 Test-Tube Burger Marks Milestone in Future Meat-Eating," *The Guardian*, Feb. 19, 2012.
74. Hanna L. Tuomisto and M. Joost Teixeira de Mattos, "Environmental Impacts of Cultured Meat Production," *Environmental Science and Technology*, July 15, 2011.

75. Gabor Forgacs, Gabor Forgacs @TEDMED Part II, May 15, 2013, http://www.youtube.com/watch?v=zDmkK8brSWk.
76. Freakonomics, "Would You Eat Steak from a Printer?" Mar. 1, 2013; link to talk by Andras Forgacs on sustainable, scalable meat, http://www.freakonomics.com/2013/03/01/would-you-eat-steak-from-a-printer/.
77. Hanna L. Tuomisto and Avijit G. Roy, "Could Cultured Meat Reduce Environmental Impact of Agriculture in Europe?" Paper presented at the 8th International Conference on LCA in the Agri-Food Sector, Rennes, France, Oct. 2–4, 2012.
78. Ketzel Levine, "Lab-Grown Meat a Reality, but Who Will Eat It?" NPR, *Morning Edition*, May 7, 2012.
79. Steven Pinker, *The Better Angels of Our Nature* (New York: Viking, 2011).

Seven
CANNIBAL COWS

1. Chengcheng Zhang et al., "Prion Protein is Expressed on Long-Term Repopulating Hematopoietic Stem Cells and Is Important for Their Self -Renewal," *Proceedings of the National Academy of Sciences* 103, no. 7 (Feb. 2006): 2184–89.
2. The first hypothesis was that a misshapen prion somehow catalyzed a normal prion, turning it into a copy of itself, with nothing but protein involved. But scientists have not yet been able to tease a concentration of pure normal prion molecules into misfolding. There are a number of other, competing theories. Laura Manuelidis, head of neuropathology at Yale University, believes an infectious slow virus might interact with protein, misfolding it. "A 25 nm Virion is the Likely Cause of Transmissible Spongiform Encephalopathies," *Journal of Cellular Biochemistry* 100, no. 4 (Mar. 2001): 897–915. See also *Yale News*, "Potentially Pathogenic Virus Found in Mad Cow Cells," Jan. 30, 2007. A group of scientists associated with the Dartmouth and Ohio State medical schools have demonstrated that a particular lipid molecule can facilitate the misfolding. Nathan R. Deleault et al., "Isolation of Phosphatidylethanolamine as a Solitary Cofactor for Prion Formation in the Absence of Nucleic Acids," *Proceedings of the National Academy of Sciences* 109, no. 22 (May 2012): 8546–51. Frank O. Bastian, clinical professor of neurosurgery and pathology and professor of veterinary science at Louisiana State University, has another theory: he builds a case that bacteria without walls, spriroplasma, are responsible for all TSEs. The biofilm that the bacteria form protects them from high heat and radiation. F. O. Bastian, "The Case for Involvement of Spiroplasma in the Pathogenesis of Transmis-

sible Spongiform Encephalopathies," *Journal of Neuropathology and Experimental Neurology* 73, no. 2 (Feb. 2014): 104–14.

3. National Institute of Neurological Disorders and Stroke, "Creutzfeldt-Jakob Disease Fact Sheet," updated December 2013. Based on their studies of the TSE disease kuru, a University College London team thinks it might take over fifty years for a few vCJD carriers to develop symptoms. John Collinge et al., "Kuru in the 21st Century—An Acquired Human Prion Disease with Very Long Incubation Periods," *The Lancet* 367, no. 9528 (June 2006): 2068–74.
4. David Brown, "The 'Recipe for Disaster' That Killed 80 and Left a £5 Bn Bill," *The Telegraph*, Oct. 27, 2000; Kamal Ahmed, Anthony Barnett, and Stuart Millar, "Madness," Special Report: The BSE Crisis, *The Guardian* and *The Observer*, Oct. 28, 2000.
5. Richard Rhodes, *Deadly Feasts* (New York: Simon & Schuster, 1998), 171–75; Sheldon Rampton and John Stauber, *Mad Cow U.S.A.* (Monroe, ME: Common Courage Press, 1997), 91.
6. Ahmed, Barnett, and Millar, "Madness."
7. Beef cattle may be fed a protein supplement during the "finishing" phase, but BSE has a long incubation period, so beef cattle didn't show symptoms before they were slaughtered. Dairy cows are often slaughtered and rendered after about three pregnancies, and their tougher meat ground up and used for hamburgers and meat pies.
8. Ahmed, Barnett, and Millar, "Madness."
9. Rampton and Stauber, *Mad Cow U.S.A.*, 93. John Wilesmith led the study.
10. Rhodes, *Deadly Feasts*, 180.
11. Department for Environment, Food, and Rural Affairs (UK), "Archive: BSE: Disease Control & Eradication—Causes of BSE," http://archive.defra.gov.uk/foodfarm/farmanimal/diseases/atoz/bse/controls-eradication/causes.htm.
12. Cornell University College of Veterinary Medicine, "Veterinary Information Brief: Mad Cow Disease and Cats," http://www.vet.cornell.edu/fhc/news/madcow.htm. No FSE cases have been reported in the United States. The Center for Food Security & Pubic Health, Iowa State University, "Feline Spongiform Encephalopathy," Sept. 2007, http://www.cfsph.iastate.edu/Factsheets/pdfs/feline_spongiform_encephalopathy.pdf.
13. Maureen Norrie, "Burger Bite," letter to the editor, *NewScientist*, Mar. 9, 2011. The whole episode is captured at: http://www.youtube.com/watch?v=QobuvWX_Grc., along with numerous other experts proclaiming with great certainty that there is no chance whatsoever of humans being infected.
14. Ahmed, Barnett, and Millar, "Madness."
15. Laura Clout, "Gummer Friend Dies of Mad Cow Disease," *The Telegraph*, Oct. 12, 2007.

16. Information on Victoria Rimmer and other early cases was derived from Rhodes, *Deadly Feasts*, and Rampton and Stauber, *Mad Cow U.S.A.* An inquest, finally performed in 2001, officially ruled that Rimmer had died of "natural causes"—that is, sporadic CJD, not variant CJD, although there was evidence of both in her brain tissue. The coroner admitted to uncertainty, and an external expert characterized her form of disease as unique. BBC News, "Inquest Uncertainty over CJD Death," Apr. 27, 2001.
17. CDC, "Fact Sheet: Variant Creutzfeldt-Jakob Disease," http://www.cdc.gov/ncidod/dvrd/vcjd/factsheet_nvcjd.htm.
18. Ahmed, Barnett, and Millar, "Madness."
19. Brown, "The 'Recipe for Disaster' That Killed 80 and Left a £5 Bn Bill."
20. Two examples are Brown Swiss cattle and Mongolian cattle. Brown Swiss may be more susceptible. C. Sauter-Louis, "Breed Predisposition for BSE: Epidemiological Evidence in Bavarian Cattle," *Schweizer Archiv für Tierheilkunde* 148, no. 5 (May 2006): 245–50. Mongolian cattle, on the other hand, might be less susceptible than some other breeds in North China. See X. Y. Zhu et al., "Bovine Spongiform Encephalopathy Associated Insertion/Deletion Polymorphisms of the Prion Protein Gene in the Four Beef Cattle Breeds from North China," *Genome* 54, no. 10 (Oct. 2011): 805–11.
21. Andy Coghlan, "Mad Cow Disease Is Almost Extinct Globally," *NewScientist*, Jan. 26, 2011; John Collinge et al., "Long Incubation Periods at the End of the Epidemic in Papua New Guinea," *Philosophical Transactions of the Royal Society* 363, no. 1510 (Nov. 2008): 3725–39.
22. Ralf F. Pettersson, presentation speech awarding the Nobel Prize in Physiology or Medicine to Stanley B. Prusiner, Dec. 10, 1997.
23. Eric M. Nicholson et al., "Identification of a Heritable Polymorphism in Bovine *PRNP* Associated with Genetic Transmissible Spongiform Encephalopathy: Evidence of Heritable BSE," *PLoS ONE* 3, no. 8 (Aug. 2008). The authors conclude: "Atypical BSE arising as both genetic and spontaneous disease, in the context of reports that at least some forms of atypical BSE can convert to classical BSE in mice, suggests a cattle origin for classical BSE." This was an H-type atypical case in the United States.
24. Sam Howe Verhovek, "Talk of the Town: Burgers v. Oprah," *New York Times*, Jan. 1, 1998.
25. CDC, "Confirmed Variant Creutzfeldt-Jakob Disease (variant CJD) Case in Texas," June 2, 2014. http://www.cdc.gov/ncidod/dvrd/vcjd/other/confirmed-case-in-texas.htm.
26. Debora MacKenzie, "American Nightmare," *NewScientist*, Aug. 7, 2004. A chilling firsthand account by slaughterman Dave Louthan, "They Are Lying About Your Food," is available at http://www.counterpunch.org/2003/01/20/they-are-lying-about-your-food. Dated Jan. 20, 2004, accessed Feb. 5, 2014.

27. Americanfarm.com, "Mad Cow Scare Cost Beef Industry Billions," May 3, 2005, http://archive.americanfarm.com/Beeftopstory5.03.05c.html.
28. ProMED Mail, BSE, Bovine—Germany (Brandenburg), "New Case, Atypical L-Type," International Society for Infectious Diseases, Jan. 17, 2014, http://www.promedmail.org/direct.php?id=2182518; World Organization for Animal Health, "Number of Reported Cases of Bovine Spongiform Encephalopathy (BSE) in Farmed Cattle Worldwide (Excluding the United Kingdom)," 2014, http://www.oie,int/?id=505.
29. Silvia Suardi et al., "Infectivity in Skeletal Muscle of Cattle with Atypical Bovine Spongiform Encephalopathy," *PLoS ONE* 7, no. 2 (Feb. 2012).
30. Kate O'Rourke, "New Form of BSE Sparks Discussion," *JAVMA News*/Ruminants, Apr. 15, 2004. The Italian researchers conducting this study found 103 cases of BSE in brain samples from over 1.6 million cows. Among the 103 cases were two samples that held the different prion.
31. Helen Thompson, "California BSE Prion Comes with a Different Twist," *Nature* NewsBlog, Apr. 27, 2012.
32. Nadine Mestre-Francés et al., "Oral Transmission of L-Type Bovine Spongiform Encephalopathy in Primate Model," Dispatch 18, no. 1 (Jan. 2012).
33. Justin J. Greenlee et al., "Clinical and Pathologic Features of H-Type Bovine Spongiform Encephalopathy Associated with E211K Prion Protein Polymorphism," *PLoS ONE* 7, no. 6 (June 2012).
34. E. M. Nicholson et al., "Identification of a Heritable Polymorphism in Bovine *PRNP* Associated with Genetic Transmissible Spongiform Encephalopathy: Evidence of Heritable BSE," *PLoS ONE* 3, no. 8 (Aug. 2008).
35. Jürgen A. Richt and S. Mark Hall, "BSE Case Associated with Prion Protein Gene Mutation," *PLoS Pathogens* 4, no. 9 (Sept. 12, 2008). Kansas State University, "Mad Cow Disease Also Caused by Genetic Mutation," *ScienceDaily*, Sept. 13, 2008, www.sciencedaily.com/releases/2008/09/080912075208.htm.
36. USDA, "Summary Report: California Spongiform Encephalopathy Case Investigation, July 2012, http://www.aphis.usda.gov/animal_health/animal_diseases/bse/downloads/BSE_Summary_Report.pdf.
37. See http://www.bseinfo.org/default.aspx, accessed Dec. 12, 2013. The "Facts You Must Know" column on the left disappeared when the page was printed. We documented it by taking a photo.
38. Patrick J. Bosque et al., "Prions in Skeletal Muscle," *Proceedings of the National Academy of Sciences* 99, no. 6 (Mar. 2002): 3812–17. Stanley B. Prusiner was a contributor to the article.
39. *Code of Federal Regulations*, Substances Prohibited from Use in Animal Food or Feed, Final Rule, title 21, sec. 589.
40. Juan Carlos Espinosa et al., "Progression of Prion Infectivity in Asymptom-

atic Cattle After Oral Bovine Spongiform Encephalopathy Challenge," *Journal of General Virology* 88, no. 4 (Apr. 2007): 1379–83.

41. Some commentators thought brain and cord matter could only be completely removed from 15 percent of cows. Others thought it could be removed from 54 percent. Revised environmental assessment (Docket. No. 2002N-0273), comment 36 at p. 22729, http://www.gpo.gov/fdsys/pkg/FR-2008-04-25/html/08-1180.htm.
42. William G. Hill, "Review of the Evidence for the Occurrence of 'BARB' BSE Cases in Cattle," July 2005, http://archive.defra.gov.uk/foodfarm/farmanimal/diseases/atoz/bse/documents/hillreport.pdf.
43. *Code of Federal Regulations*, title 21, Substances Prohibited from Use in Human Food, subpart B sec. 189.5: Prohibited Cattle Materials. The definition of SRM applies only to cattle thirty months and older. Current as of Apr. 1, 2013.
44. Veterinary Laboratories Agency (now merged with Animal Health to form the Animal Health and Veterinary Laboratories Agency of the British Ministry of Agriculture, Fisheries and Food), "Information Released on 2 February 2005," http://vla.defra.gov.uk/vla/vla_ati_020205.htm, accessed Nov. 19, 2012.
45. Michael Hansen, "Consumers Union's Comments on FDA Docket No. 2002N-0273: Substances Prohibited from Use in Animal Food and Feed, Dec. 20, 2005, 2.
46. Environmental Assessment for Amendments to *Code of Federal Regulations*, Substances Prohibited from Use in Animal Food or Feed, title 21, sec. 589, Final Rule, Apr. 2008.
47. The FDA is only responsible for finding unsafe dietary supplements *after* they are for sale, so products like Brain 360, made by Atrium Nutrition, are still being sold. Brain 360 advertises that it is made from "Raw Brain Tissue Concentrate (not an extract) of bovine source. Tissues processed by Low Temperature Lyophilization by Sublimination [*sic*] to insure rawness and preserve natural constituents," Bayho.com, http://www.bayho.com/p/837025.html, accessed Nov. 13, 2012.
48. USDA BSE (Mad Cow Disease) Ongoing Surveillance Information Center.
49. USDA Office of Inspector General, "Audit Report, Animal and Plant Health Inspection Service, Bovine Spongiform Encephalopathy (BSE) Surveillance Program—Phase II, and Food Safety and Inspection Service Controls Over BSE Sampling, Specified Risk Materials, and Advanced Meat Recovery Products—Phase III," Report No. 50601-10-KC, Jan. 2006, http://www.usda.gov/oig/rptsauditsfsis.htm.
50. Donald G. McNeil, Jr., "U.S. Reduces Testing for Mad Cow Disease, Cit-

ing Few Infections," *New York Times*, Dec. 1, 2004: "Testing is voluntary, and the department pays about $100 for samples, so sampling was not random. Slaughterhouses eager to recoup some of their disposal costs for dead animals, but not eager to be shut down, had an incentive to send in samples from animals less likely to test positive"; MacKenzie, "American Nightmare."

51. Marc Kaufman, "Company's Mad Cow Tests Blocked; USDA Fears Other Firms' Meat Would Appear Unsafe," *Washington Post*, Apr. 16, 2004.
52. Michael Hansen, "Stop the Madness," *New York Times* op-ed, June 20, 2008. Virginie Supervie of the University of Paris VI, an epidemiologist, has done calculations for France that suggest similar dramatic undercounting there. Virginie Supervie and D. Costagliola, "The Unrecognized French BSE Epidemic," *Veterinary Research* 35, no. 3 (May–June 2004): 349–62. Britain's cow population was about 10 million in 2010, http://archive.defra.gov.uk/foodfarm/farmanimal/diseases/vetsurveillance/species/cattle/.
53. Jonathan Turley, "The Beef People: The Creekstone Controversy and the Bush Administration's Effort to Prevent Private Testing of Meat Products," Nov. 15, 2007, http://jonathanturley.org/2007/11/15/the-beef-people-creekstone-and-the-bush-administrations-effort-to-prevent-private-testing-of-meat-products/. First published by the *Los Angeles Times*, Apr. 20, 2004.
54. *Creekstone Farms Premium Beef, L. L. C. v. Department of Agriculture*, 517 F. Supp.2d 8 (D.D.C. Cir. 2007); *Creekstone Farms Premium Beef, L.L.C. v. Department of Agriculture*, 539 F. 3d 492 (D.C. Cir. 2008).
55. USDA FSIS, "Bovine Spongiform Encephalopathy—'Mad Cow Disease,'" http://www.fsis.usda.gov/wps/portal/fsis/topics/food-safety-education/get-answers/food-safety-fact-sheets/production-and-inspection/bovine-spongiform-encephalopathy-mad-cow-disease/bse-mad-cow-disease.
56. *New York Times* editorial board, "The Biggest Beef Recall Ever," *New York Times*, Feb. 21, 2008.
57. Andrew Martin, "Humane Society Criticized in Meat Quality Scandal," *New York Times*, Feb. 27, 2008. To view the parts of the video shown on prime-time television: http://www.cbsnews.com/stories/2008/01/30/eveningnews/main3773183.shtml.
58. Victoria Kim, Mitchell Landsberg, "Huge Beef Recall," *Los Angeles Times*, Feb. 18, 2008.
59. Electroencephalograms can reveal brainwave patterns associated with CJD or vCJD, MRIs might also be helpful, and ways to test cerebrospinal fluid for protein markers for CJD are under development. Tonsil biopsies are more helpful in finding vCJD than in diagnosing CJD. One or more types of blood test for prion diseases should soon be available; news reports say

one such test is already being used in the UK. Victoria MacDonald "Blood Test Breakthrough for 'Mad Cow Disease,'" 4 News (London), Jan. 13, 2012. A saliva test developed by the National Institute of Animal Health in Japan was recently used to detect infectious prions in cows' spit before, or just as, symptoms appeared.

60. CDC, *MMWR Recommendations and Reports*, "Mandatory Reporting of Infectious Diseases by Clinicians," June 22, 1990, http://www.cdc.gov/mmwr/preview/mmwrhtml/00001665.htm.
61. Washington State Department of Health, "Prion Disease," http://www.doh.wa.gov/YouandYourFamily/IllnessandDisease/PrionDisease.aspx.
62. Stanley B. Prusiner, "A Unifying Role for Prions in Neurodegenerative Diseases," *Science* 336, no. 6088 (June 2012): 1511–13; Laura Sanders, "Prions May Cause Alzheimer's: Similarity Found with Destructive Protein Behind Mad Cow," *Science News* 182, no. 13 (Dec. 2012). Might Alzheimer's also be infectious? Nearly all doctors currently believe Alzheimer's is not infectious. But hairline cracks are appearing in this belief. Scientists at the University of Texas Health Science Center took material from the brain of a confirmed human Alzheimer patient and injected it into mice. All the injected mice developed plaques and other changes in their brains of the sort seen in Alzheimer's patients. See "Alzheimer's Disease Transmission May Be Similar to Infectious Prion Diseases," University of Texas, news release, Oct. 4, 2011, http://www.uth.edu/media/story.htm?id=3541058. A scientist on the Texas team, Claudio Soto, said, "Our findings open the possibility that some of the sporadic Alzheimer's cases may arise from an infectious process, which occurs with other neurological diseases such as mad cow and its human form, Creutzfeldt-Jakob disease."

    Different scientists took samples of amyloid-beta from the brains of Alzheimer's patients and injected the samples into marmoset monkeys and mice genetically engineered to be susceptible to the disease. The injections triggered Alzheimer's in the animals. See Roxanne Khamsi, "Alzheimer's May 'Seed' Itself Like Mad Cow Disease," *NewScientist*, Sept. 21, 2006; R. Morales et al., "De Novo Induction of Amyloid-β Deposition in Vivo," *Molecular Psychiatry* 17, no. 12 (Dec. 2012): 1347–53.
63. E. E. Manuelidis and L. Manuelidis, "Suggested Links Between Different Types of Dementias: Creutzfeldt-Jakob Disease, Alzheimer Disease, and Retroviral CNS Infections," *Alzheimer Disease and Associated Disorders* 3, nos. 1–2 (Spring–Summer 1989): 100–9. "In our own neuropathological material, in 46 cases diagnosed clinically as AD [Alzheimer's Disease], 6 cases were proven to be CJD at autopsy [13 percent]."
64. The brains of fifty-four "dementia" patients were examined at the University

of Pittsburgh, and three of the brains (5.5 percent) were found to be infected with CJD. F. Boller, O. L. Lopez, and J. Moossy, "Diagnosis of Dementia: Clinicopathologic Correlations, *Neurology* 39, no. 1 (Jan. 1989): 76–79. Other studies found about a 3 percent rate—see Bala Mahendra, *Dementia: A Survey of the Syndrome of Dementia* (Lancaster, England: MTP Press Limited, 1987), 174; J. P. H. Wade et al., "The Clinical Diagnosis of Alzheimer's Disease," *Archives of Neurology* 44 (1987): 24–29.

65. See http://www.cjdsurveillance.com/abouthpd-animal.html.
66. Jane Zhang, "U.S. News: Meat Inspectors Can't Keep Up, Official Says," *Wall Street Journal* (Eastern edition), Apr. 18, 2008.
67. Marc Kaufman, "Agency Fought Retesting of Infected Cow," *Washington Post*, Feb. 3, 2006.
68. Erica Goode, "Fewer Cows' Hides May Bear the Mark of Home," *New York Times*, Jan. 26, 2012; Lisa M. Krieger, "Cattle Branding May Be Put Out to Pasture," *San Jose Mercury News*, reprinted in the *Seattle Times*, Jan. 17, 2012.
69. Verlyn Klinkenborg, "The Whole Cow and Nothing but the Whole Cow," *New York Times*, Jan. 20, 2004.
70. USDA "Animal Disease Traceability Framework," updated Aug. 14, 2013, http://www.aphis.usda.gov/traceability/downloads/ADT_eartags_criteria.pdf. Link to Final Rule is on this site, at http://www.aphis.usda.gov/traceability/.
71. K. C. Jones, "Invisible RFID Ink Safe for Cattle and People Company Says," *Information Week*, Jan. 10, 2007, http://www.informationweek.com/invisible-rfid-ink-safe-for-cattle-and-people-company-says/d/d-id/1050602?.
72. FDA, "Environmental Assessment for Amendments to 21 CFR 589.2001, Substances Prohibited from Use in Animal Food or Feed to Prevent the Transmission of Bovine Spongiform Encephalopathy, Final Rule," http://www.fda.gov/ohrms/dockets/98fr/fda-2002-n-0031-ea.pdf.
73. Clell V. Bagley, John H. Kirk, Kitt Farrell-Poe, "Cow Mortality Disposal," Utah State University Extension, Oct. 1999, http://extension.usu.edu/files/publications/publication/AG-507.pdf.
74. Christopher J. Johnson et al., "Prions Adhere to Soil Minerals and Remain Infectious," *PLoS Pathogens* 2, no. 4 (Apr. 2006); P. Brown and D. C. Gajdusek, "Survival of Scrapie Virus After 3 Years' Interment," *The Lancet* 2, no. 337 (1991): 269–70. Richard C. Wiggins, "Prion Stability and Infectivity in the Environment," *Neurochemical Research* 34, no. 1 (Jan. 2009): 158–68.
75. C. Bartholomay et al., "The Fate of Infectious Prion Proteins in Wastewater Treatment Systems," *Proceedings of the Water Environment Federation*,

WEFTEC 2005: Sessions 91–100, 8116–8118(3). However, water treatment systems that use alkaline hydrolysis (addition of lime) should inactivate prions.

76. CDC, *Biosafety in Microbiological Laboratories (BMBL)*, 5th ed. (Washington, D.C.: Government Printing Office, Dec. 2009), section VIII-H: Prion Diseases, 285–86.
77. H. Leon Thacker, *Carcass Disposal: A Comprehensive Review*, chap. 6 "Alkaline Hydrolysis," National Agricultural Biosecurity Center Consortium, USDA APHIS Cooperative Agreement Project, Carcass Disposal Working Group, Aug. 2004.
78. BioLiquidator; Alkaline Hydrolysis System, http://www.bioliquidator.com/.
79. Don A. Franco, "Animal Disposal—The Environmental, Animal Disease, and Public Health Related Implications: An Assessment of Option," presentation to the California Department of Food and Agricultural Symposium, Sacramento, CA, Apr. 8, 2002. Reprinted in *Render*, http://rendermagazine.com/industry/animal-disposal/, accessed Nov. 20, 2012.
80. James McWilliams, "The Deadstock Dilemma: Our Toxic Meat Waste, *The Atlantic*, Sept. 11, 2010.
81. Darling International, "Darling International Inc. to Acquire Griffin Industries in Transaction Valued at Approximately $840 Million," news release, Nov. 9, 2010. http://ir.darlingii.com/profiles/investor/ResLibraryView.asp?ResLibraryID=41590&GoTopage=4&Category=1191&BzID=1640, accessed Aug. 14, 2013; *Render*, "Darling International Acquires Griffin Industries," Dec. 10, 2010.
82. Coghlan, "Mad Cow Disease Is Almost Extinct Globally."

Eight
WHY BUY ORGANIC BEEF AND DAIRY?

1. Carol Pogash, "The Elders of Organic Farming," *New York Times*, Jan. 25, 2014.
2. Tom Forsythe, "The One and Only Cheerios," Jan. 2, 2014, http://www.blog.generalmills.com/2014/01/the-one-and-only-cheerios/.
3. General Mills, "State-Based Labeling Laws: Washington Initiative 522," undated,http://www.generalmills.com/ChannelG/Issues/state_based_labeling.aspx.
4. The nonprofit, independent Organic Materials Review Institute has a downloadable product list containing over 2,800 "OMRI Listed®" products that have been independently screened and tested by the institute for compliance

withfederalstandards,http://www.omri.org/sites/default/files/opl_pdf/crops_category.pdf.

5. USDA, "National Organic Standards Board," updated Sept. 16, 2013, http://www.ams.usda.gov/amsv1.0/nosb.
6. April Fulton, "Whole Foods Founder John Mackey on Fascism and 'Conscious Capitalism,'" NPR, *Salt Blog*, Jan. 16, 2013; Josh Harkinson,"Whole Foods CEO Welcomes Climate Change, Warns of Fascism," *Mother Jones*, Jan.18, 2013.
7. Phil Howard, "Organic Industry Structure," The Cornucopia Institute, http://www.cornucopia.org/dairysurvey/.
8. Cornucopia Institute, "Organic Dairy Report/Ratings," http://www.cornucopia.org/dairysurvey/. Whole Foods was the only store-brand milk to cooperate with the Cornucopia survey.
9. Cliff Feigenbaum, "Interview: George Siemon, CEO of Organic Valley," *Green Money Journal*, Sept. 5, 2012. In an industry where prices can be volatile and where organic prices tend to be more volatile than those for conventional milk, unpredictable prices were driving small dairy farms out of business. By gathering them into a huge national coop (more than 1,800 dairies, about half of them with fifty or fewer cows), Siemon and his team were able to bring stability to many organic farmers.
10. Melanie Warner, "Wal-Mart Eyes Organic Foods," *New York Times*, May 12, 2006.
11. Tom Philpott, "Is Walmart Really Going Organic and Local?" *Mother Jones*, Mar.–Apr. 2012.
12. This story was broken by an electronic news service, consulting, and conference firm, greenbiz.com, with an excellent, balanced three-part series: Joel Makower, "Exclusive: Inside McDonald's Quest for Sustainable Beef," "How a Big Mac Becomes Sustainable," and "Can the Beef Industry Collaborate Its Way to Sustainability?" Jan. 7–9, 2014.
13. Vanessa Wong, "Fleshing Out the Incredibly Vague Concept of 'Sustainable Beef,'" *Bloomberg Businessweek*, Mar. 20, 2014.
14. James Boswell, *Boswell's Life of Johnson*, abridged and edited by Charles Grosvenor Osgood, 2013, http//www.gutenberg.org/files/1564-h/1564-h.htm.
15. Rebecca Lindsey, "Looking for Lawns," NASA/Earth Observatory, discussing the work of Cristina Milesi and her team at NASA's Ames Research Center.
16. Ronald M. Nowak, *Walker's Mammals of the World* (Baltimore: Johns Hopkins University Press, 1999).
17. Thomas A. Arcury et al., "Farmworker Exposure to Pesticides: Methodologic Issues for the Collection of Comparable Data," *Environmental Health Perspectives* 114, no. 6 (June 2006): 923–28.

18. E. Palupi et al., "Comparison of Nutritional Quality Between Conventional and Organic Dairy Products: A Meta-Analysis," *Journal of the Science of Food and Agriculture* 92, no. 14 (Nov. 2012): 2774–81.
19. S. K. Duckett et al., "Effects of Winter Stocker Growth Rate and Finishing System on: III. Tissue Proximate, Fatty Acid, Vitamin, and Cholesterol Content," *Journal of Animal Science* 87, no. 9 (Sept. 2009): 2961–70.
20. Cynthia A. Daley et al., "A Review of Fatty Acid Profiles and Antioxidant Content in Grass-Fed and Grain-Fed Beef," *Nutrition Journal* 9, no. 10 (Mar. 2010), http://www.nutritionj.com/content/9/1/10; Tom Philpott, "UK Organic Milk Better for You Than Conventional, Thanks to Cows' Grass-Based Diet," *Grist*, Jan. 20, 2011. Organic milk provided around 40 percent more CLA than conventional milk, and summer milk (when grass is freshest) was better than winter milk. Conventional dairy cows get more pasture time in England than they do in the United States, so there is probably an even greater difference in America between the CLA content of conventional milk and that of organic milk.
21. T. R. Dhiman, "Conjugated Linoleic Acid: A Food for Cancer Prevention," Proceedings from the 2000 Intermountain Nutrition Conference, 103–21, cited on Eatwild.com, "Health Benefits of Grass-Fed Products," http://www.eatwild.com/healthbenefits.htm.
22. L. Rist et al., "Influence of Organic Diet on the Amount of Conjugated Linoleic Acids in Breast Milk of Lactating Women in the Netherlands," *British Journal of Nutrition* 97, no. 4 (Apr. 2007): 735–43.
23. Duckett et al., "Effects of Winter Stocker Growth Rate and Finishing System on: III. Tissue Proximate, Fatty Acid, Vitamin, and Cholesterol Content." See also eatwild.com, which has links to papers on the health benefits of grass-fed products, including meat, dairy, and eggs; Daley et al., "A Review of Fatty Acid Profiles and Antioxidant Content in Grass-Fed and Grain-Fed Beef"; Gillian Butler et al., "Fatty Acid and Fat-Soluble Antioxidant Concentrations in Milk from High- and Low-Input Conventional and Organic Systems: Seasonal Variation," *Journal of the Science of Food and Agriculture* 88 (2008): 1431–41.
24. Clancy, *Greener Pastures*, 1, 40.
25. Daley et al., "A Review of Fatty Acid Profiles and Antioxidant Content in Grass-Fed and Grain-Fed Beef"; J. M. Leheska et al., "Effects of Conventional and Grass Feeding Systems on the Nutrient Composition of Beef," *Journal of Animal Science* 86, no. 12 (Dec. 2008): 3575–85.
26. Palupi et al., "Comparison of Nutritional Quality Between Conventional and Organic Dairy Products: A Meta-Analysis"; Gillian Butler et al., "Fat Composition of Organic and Conventional Retail Milk in Northeast

England," *Journal of Dairy Science* 94, no. 1 (Jan. 2011): 24–36. Summer milk had lower concentrations of saturated fatty acids and higher concentrations of the good acids than did winter milk. But summer or winter, organic milk had more beneficial fats. Whether it was being raised in pasture or raised organically that made the difference wasn't determined. See also K. A. Ellis et al., "Comparing the Fatty Acid Composition of Organic and Conventional Milk," *Journal of Dairy Science* 89, no. 6 (June 2006): 1938–50.

27. A. P. Simopoulos, "The Importance of the Ratio of Omega-6/Omega-3 Essential Fatty Acids," *Biomedicine & Pharmacotherapy* 56, no. 8 (Oct. 2002): 365–79.
28. T. A. Dolecek and G. Grandits, "Dietary Polyunsaturated Fatty Acids and Mortality in the Multiple Risk Factor Intervention Trial (MRFIT)," *World Review of Nutrition and Dietetics* 66 (1991): 205–16.
29. FDA, "FDA Announces Qualified Health Claims for Omega-3 Fatty Acids," news release, Sep. 8, 2004. The FDA does not require labels to specify what type of omega-3s are in a product, so a label saying a product contains "omega-3" doesn't mean that it contains EPA or DHA.
30. Canadian Food Inspection Agency, "Chapter 8; Health Claims; Sections 8.1–8.6, http://www.inspection.gc.ca/english/fssa/labeti/guide/ch8e.shtml#a8_2, Table 8-3: "Summary Table of Acceptable Nutrient Function Claims." A footnote says: "This claim is based on available scientific evidence indicating that the development of the brain, eyes, and nerves in the human infant takes place very early starting in late pregnancy and up to 2 years of age. The Institute of Medicine in their 2005 report [*Dietary Reference Intakes for Energy, Carbohydrate, Fiber, Fat, Fatty Acids, Cholesterol, Protein, and Amino Acids* (Washington, D.C.: The National Academies Press, 2005), 444–45] stated that 'The developing brain accumulates large amounts of DHA during the pre- and postnatal development and this accumulation continues throughout the first 2 years after birth.'"
31. Perry Brewer and Chris R. Calkins, "Quality Traits of Grain- and Grass-Fed Beef: A Review," University of Nebraska (Lincoln), Digital Commons, Jan. 1, 2003; J. Severe and D. R. ZoBell, "Grass-Fed vs. Conventionally Fed Beef," Utah State University Cooperative Extension, May 2011. Both these studies concluded that grass-fed was less tender. But not all experts agree. J. D. Crouse, H. R. Cross, and S. C. Seideman, "Effects of a Grass or Grain Diet on the Quality of Three Beef Muscles," *Journal of Animal Science* 58, no. 3 (Mar. 1984): 619–25. According to Fred Martz, "Pasture-Based Finishing of Cattle and Eating Quality of Beef," University of Missouri Forage Systems Research Center, Mar. 21, 2000, "Aging for three weeks greatly improved tenderness,

especially of pasture-based finished beef, without greatly affecting meat flavor or aroma. The results of these trials support the concept that cattle can be finished on pasture to meet the demands of the conventional beef trade or can be finished without grain supplementation to produce smaller, leaner carcasses."

32. Chunping Zhao et al., "Functional Genomic Analysis of Variation on Beef Tenderness Induced by Acute Stress in Angus Cattle," *Comparative and Functional Genomics* (2012).
33. Kathryn Shattuck, "Where Corn is King, a New Regard for Grass-Fed Beef," *New York Times*, June 17, 2013.
34. *Federal Register*, "Grass (Forage) Federal Marketing Claim Standard," vol. 72, no. 199 (Oct. 16, 2007): 58631, http://www.ams.usda.gov/amsv1.0/getfile?dDocName=STELPRDC5063842.
35. Pesticide Action Network, What's on My Food?/Beef Fat, "10 Pesticide Residues Found by the USDA Pesticide Data Program," http://www.whatsonmyfood.org/food.jsp?food=BA.
36. M. F. Bouchard, "Prenatal Exposure to Organophosphate Pesticides and IQ in 7-Year-Old Children," *Environmental Health Perspectives* 119, no. 8 (Aug. 2011): 1189–95.
37. Chensheng Lu et al., "Organic Diets Significantly Lower Children's Dietary Exposure to Organophosphorus Pesticides," *Environmental Health Perspectives* 114, no. 2 (Feb. 2006): 260–63.
38. Jamie Hirsh, "Don't Give Up on Organic Food, Our Experts Urge," *Consumer News*, Sept. 5, 2012, http://news.consumerreports.org/health/2012/09/dont-give-up-on-organic-food-our-experts-urge.html. Charles Benbrook, of the Center for Sustaining Agriculture and Natural Resources at Washington State University, says that there is "encouraging evidence that organic food can reduce the odds of some adverse health impacts, including birth defects, neuro-behavioral and learning problems, autism, and eczema." Benbrook notes that there is now "strong evidence" that prenatal exposures to organophosphate insecticides might lower a baby's IQ: Charles Benbrook, "Initial Reflections on the *Annals of Internal Medicine* paper 'Are Organic Foods Safer and Healthier Than Conventional Alternatives? A Systematic Review," Sept. 4, 2012, http://caff.org/wp-content/uploads/2010/07/Annals_Response_Final.pdf.
39. American Academy of Pediatrics, "AAP Makes Recommendations to Reduce Children's Exposure to Pesticides," press release, Nov. 26, 2012, http://www.aap.org/en-us/about-the-aap/aap-press-room/Pages/AAP-Makes-Recommendations-to-Reduce-Children's-Exposure-to-Pesticides.aspx.
40. Robert Pear, "Some Stores Cater to Poor but Bill U.S for Top Prices," *New York Times*, June 6, 2004. Another problem is the growth of WIC specialty

stores, which sell only approved foods but charge much more. WIC customers don't care, because their vouchers buy a set "package" of foods, paid for by federal grants.

41. USDA, "National Organic Program," http://www.ams.usda.gov/AMSv1.0/nop.
42. Brian P. Baker et al., "Pesticide Residues in Conventional, IPM-Grown and Organic Foods: Insights from Three U.S. Data Sheets," *Food Additives and Contaminants* 19, no. 5 (May 2002): 427–46.
43. NIH/National Institute of Environmental Health Sciences, "Two Pesticides—Rotenone and Paraquat—Linked to Parkinson's Disease, Study Suggests," *ScienceDaily*, Feb. 15, 2011, http://www.sciencedaily.com/releases/2011/02/110214115442.htm.
44. Christine A. Bahlai et al., "Choosing Organic Pesticides over Synthetic Pesticides May Not Effectively Mitigate Environmental Risk in Soybeans," *PLoS ONE* 5, no. 6 (June 2010), http://www.plosone.org/article/info%3Adoi%2F10.1371%2Fjournal.pone.0011250.
45. Laura Pickett Pottorff, "Some Pesticides Permitted in Organic Gardening," Colorado State Cooperative Extension, http://www.colostate.edu/Depts/CoopExt/4DMG/VegFruit/organic.htm.
46. K. Hayden et al., "Occupational Exposure to Pesticides Increases the Risk of Incident AD: The Cache County Study," *Neurology* 74, no. 9 (May 2010): 1524–30.
47. USDA/Economic Research Service, "Adoption of Genetically Engineered Crops in the U.S.: Recent Trends in GE Adoption," July 9, 2013, http://www.ers.usda.gov/data-products/adoption-of-genetically-engineered-crops-in-the-us/recent-trends-in-ge-adoption.aspx#.U4yLE_ldXa4.
48. Washington State University, "'Superweeds' Linked to Rising Herbicide Use in GM Crops, Study Finds," *ScienceDaily*, Oct. 2, 2012—story based on materials provided by Washington State University in an article originally written by Brian Clark; Charles M. Benbrook, "Impacts of Genetically Engineered Crops on Pesticide Use in the U.S.—The First Sixteen Years," *Environmental Sciences Europe*, 2012.
49. Monsanto, "Agricultural Herbicides," http://www.monsanto.com/products/Pages/agricultural-herbicides.aspx, accessed Sept. 11, 1013. Monsanto's patent expired in 2000, and many other firms now also make glyphosate products.
50. Andrew Pollack, "E.P.A. Denies an Environmental Group's Request to Ban a Widely Used Weed Killer," *New York Times*, Apr. 9, 2012.
51. Wissem Mnif et al., "Effect of Endocrine Disruptor Pesticides: A Review," *International Journal of Environmental Research and Public Health* 8, no. 6 (June 2011): 2265–2303.

52. D. C. Jones and G. W. Miller, "The Effects of Environmental Neurotoxicants on the Dopaminergic System: a Possible Role in Drug Addiction," *Biochemical Pharmacology* 76, no. 5 (Sept. 2008): 569–81.
53. Andrew Pollack, "Dow Corn, Resistant to a Weed Killer, Runs into Opposition," *New York Times*, Apr. 25, 2012. Dow is working on ways to make 2,4-D less likely to vaporize and drift.
54. Charles M. Benbrook, "Impacts of Genetically Engineered Crops on Pesticide Use in the U.S.—The First Sixteen Years," *Environmental Sciences Europe* 24, no. 24 (2012).
55. Avik Mukherjee et al., "Preharvest Evaluation of Coliforms, *Escherichia coli*, Salmonella, and *Escherichia coli* O157:H7 in Organic and Conventional Produce Grown by Minnesota Farmers," *Journal of Food Protection* 67, no. 5 (2004): 894–900. The Bullitt Center building in Seattle where co-author Denis works is the only six-story office building in the world with composting toilets. It processes its compost for eighteen months. And no, the toilets don't stink.
56. Adam S. Davis et al., "Increasing Cropping System Diversity Balances Productivity, Profitability and Environmental Health," *PLoS ONE* 7, no. 10 (Oct. 2012).
57. Nathanael Johnson, "Organic Farming Sucks (Up Carbon)," *Grist*, June 12, 2013; A. Gattinger et al., "Enhanced Top Soil Carbon Stocks under Organic Farming," *Proceedings of the National Academy of Sciences*, Oct. 30, 2012, http://www.pnas.org/content/109/44/18226.full.
58. *Code of Federal Regulations*, "Livestock Living Conditions," title 7, sec. 205.239(d).

Nine
COWBOYS VS. ASTRONAUTS

1. Frederick Jackson Turner, *The Frontier in American History* (New York: Henry Holt and Company, 1921).
2. Interestingly, the first economic reference to the earth as a spaceship described it from the cornucopian perspective. In *Progress and Poverty*, Henry George wrote, "It is a well-provisioned ship, this on which we sail through space. If the bread and beef above decks seem to grow scarce, we but open a hatch and there is a new supply, of which before we never dreamed. And very great command over the services of others comes to those who as the hatches are opened are permitted to say, 'This is mine.'" George was living through the disappearance of the frontier and the enclosure of land,

and was famous for developing a radical, creative tax proposal to address the resulting distributional ills. Henry George, *Progress and Poverty* (New York: Modern Library, 1947), 204.

3. Adlai Stevenson, speech to the Economic and Social Council of the United Nations, Geneva, Switzerland, July 9, 1965.
4. Arguably, this perspective was first offered by the distinguished historian David Potter, in *People of Plenty: Economic Abundance and the American Character* (Chicago: University of Chicago Press, 1958). The viewpoint was transformed into ideological armor by right-wing think tanks that supported big business against environmental constraints. Now it is increasingly identified with the one-tenth of 1 percent of Silicon Valley to whom the technological revolution has been most generous. Its most sophisticated treatment to date is in Peter Diamandis and Steven Kotler, *Abundance: the Future Is Better Than You Think* (New York: Free Press, 2012).
5. A readable summary of past farm bills can be found in Scott Marlow's *The Non-Wonk Guide to Understanding Federal Commodity Payments* (Pittsboro, NC: The Rural Advancement Foundation International—USA, 2005), http://www.rafiusa.org/pubs/nonwonkguide.
6. *Wall Street Journal*, "The Farm State Pig-Out," editorial, May 2, 2002.
7. Associated Press, "Making Corn-Based Ethanol Badly Hurting Environment," cbsnews.com, Nov. 12, 2013.
8. In a bit of clever legislative legerdemain, the Farm Bill combines agricultural programs designed to gather support from Big Ag with food assistance for the poor (now called the Supplemental Nutrition Assistance Program, or SNAP). SNAP is favored by urban liberals. With SNAP payments of about $75 billion in 2013 flowing to about one out of every seven Americans, Tea Party hostility boiled over, and the Farm Bill coalition unraveled. House Republicans argued that SNAP was out of control. Even as they boosted farm subsidies, they cut SNAP by $40 billion. Democrats responded that two-thirds of SNAP beneficiaries were children, the elderly, or disabled, and that most of the rest were adults who were caring for children. As it was designed to do, SNAP had grown during the financial recession of 2008 and the high-unemployment recovery that followed. The bill passed without a single Democratic vote, in the face of a White House veto threat, and with warnings that it would be dead on arrival in the Senate. House Budget Committee chair Paul Ryan warned against turning "the safety net into a hammock that lulls able-bodied people to lives of dependency and complacency." Economist Paul Krugman, noting that average SNAP benefits are $4.45 a day, commented trenchantly, "some hammock" (Krugman, "Free to be Hungry," *New York Times*, Sept. 22, 2013). See also the Center on Bud-

get and Policy Priorities, "SNAP Enrollment Remains High Because the Job Market Remains Weak, http://www.cbpp.org/research/?fa=topic&id=69.

9. *St. Louis Post-Dispatch* editorial staff (rerun in the *Globe Gazette* [Mason City, Iowa]), "Farm Bill Sends Our Country in Wrong Direction," *St. Louis Post-Dispatch*, Feb. 5, 2014.
10. *Des Moines Register* editorial staff, "Deal on Farm Bill Leaves Many Disappointed," *Des Moines Register*, Jan. 30, 2014.
11. 42 U.S.C. §§4321–4370e (2006).
12. Worldwatch Institute, "Agribusinesses Consolidate Power," product number VST028.
13. *New York Times* editorial staff, "Reforming Meat," *New York Times*, Sept. 7, 2010.
14. John McCormick et al., "Farmers' Penalties Rarely Stick," *Des Moines Register*, Apr. 20, 2002.
15. USDA Center for Nutrition Policy and Promotion, *Dietary Guidelines for Americans, 2010*, 46, http://www.health.gov/dietaryguidelines/2010.asp.
16. CDC, National Center for Chronic Disease Prevention and Health Promotion, "State Indicator Report on Fruits and Vegetables," 2013.
17. Matt Milkovich, "Growers Need Crop Insurance to Be More Transparent," *Fruit Growers News*, July 2, 2012, http://fruitgrowersnews.com/index.php/magazine/article/growers-need-crop-insurance-to-be-more-transparent; Michael Moss, "The Seeds of a New Generation," *New York Times*, Feb. 4, 2014.
18. Fernando Gomez-Pinilla and Rahul Agrawal, "'Metabolic Syndrome' in the Brain: Deficiency in Omega-3 Fatty Acid Exacerbates Dysfunctions in Insulin Receptor Signalling and Cognition," *Journal of Physiology* 590, no. 1 (May 2012). Summarized in *ScienceDaily*, "This is Your Brain on Sugar: Study Shows High-Fructose Diet Sabotages Learning, Memory," May 15, 2012. Adding omega-3 fatty acids to the rats' diets protected the rats from brain dysfunction. The Corn Refiners Association responded: "There is abundant scientific evidence demonstrating that consuming fructose and glucose together is entirely safe due to the way the body metabolizes these simple sugars in combination," http://sweetsurprise.com/press/response-ucla-rat-study.
19. USDA Economic Research Service, "Percent of Household Final Consumption Expenditures Spent on Food, Alcoholic Beverages, and Tobacco That Were Consumed at Home, by Selected Countries, 2012," updated Sept. 16, 2013. See also "Clear These Healthy-Eating Hurdles," *Consumer Reports on Health*, Aug. 2012, 10, which recommends nutrient-rich but inexpensive foods like barley, quinoa, beans, eggs, and fresh produce that's in season.

20. Doug Gurian-Sherman, *CAFOs Uncovered: The Untold Costs of Confined Animal Feeding Operations* (Cambridge, MA: Union of Concerned Scientists, 2008), 6, table ES-1. This study covers hog and poultry CAFOs as well as dairy and beef CAFOs.
21. Union of Concerned Scientists, "The Hidden Costs of CAFOs," issue briefing, Sept. 2008, http://www.ucsusa.org/assets/documents/food_and_agriculture/cafo_issue-briefing-low-res.pdf.
22. Shane Ellis, "State of the Beef Industry 2009," Iowa State University, 2009, http://beefmagazine.com/site-files/beefmagazine.com/files/archive/beefmagazine.com/BEEF_SOI_2009.pdf.
23. Tracie McMillan, "Why Your Hamburger Hates America," *Washington Post*, June 29, 2010.
24. Frank Morris, "Antitrust Official Gets Stampeded by Big Beef," NPR, *The Salt*, Jan. 25, 2012.
25. The National Forest Service (part of the USDA) manages 193 million acres, and the Bureau of Land Management (in the Department of Interior) operates an additional 247 million acres.
26. Pulitzer Prize–winning reporter Tom Knudson wrote a superb three-part series on this topic in 2012 for the *Sacramento Bee:* "The Killing Agency: Wildlife Services' Brutal Methods Leave a Trail of Animal Death," Apr. 29, 2012; "Wildlife Services' Deadly Force Opens Pandora's Box of Environmental Problems," Apr. 30, 2012; "Suggestions in Changing Wildlife Services Range from New Practices to Outright Bans," May 6, 2012. NRDC estimates that Wildlife Services kills more than one million animals each year.
27. Mark Salvo, "Mortgaging Public Assets: How Ranchers Use Grazing Permits as Collateral," in *Welfare Ranching*, edited by Wuerthner and Matteson, 271–73.
28. Thomas M. Power, "Taking Stock of Public Lands Grazing," in *Welfare Ranching*, edited by Wuerthner and Matteson, 263–69.
29. Ibid.
30. Deborah Epstein Popper and Frank J. Popper, "The Great Plains: From Dust to Dust," *Planning*, Dec. 1987, http://www.csupomona.edu/~tgyoung/rs510/Grt_Plains.pdf.
31. Of the roughly 500,000 bison now alive in America, virtually all are actually hybrids of bison and cattle. Fewer than 7,000 are pureblood bison.
32. American Prairie, Frequently Asked Question," http://www.americanprairie.org/aboutapf/faqs/.
33. Jack Healy, "Vision of Prairie Paradise Troubles Some Montana Ranchers," *New York Times,* Oct. 27, 2013; *The Economist*, "Born to Be Wild: Buffalo Are Coming Back to the American Prairie," Mar. 17, 2012.

34. Stephanie Strom, “Parched in Cattle Country: A Long Drought Tests Ranchers’ Patience and Creativity,” *New York Times*, Apr. 6, 2013.
35. Office of Management and Budget, “Dairy Price Support Program,” 2006, http://georgewbushwhitehouse.archives.gov/omb/expectmore/summary/1000 2436.2006.html.
36. Tom Webb, “Federal Policy Creates Milk Mountains,” Aug. 17, 2003, http:// www.highbeam.com/doc/1G1-106713275.html.
37. Gilbert M. Gaul et al., “Aid to Ranchers Was Diverted for Big Profits at Taxpayers’ Expense,” *Washington Post*, July 19, 2006.
38. Dennis A. Shields, “Consolidation and Concentration in the U.S. Dairy Industry,” Congressional Research Service, Apr. 27, 2010, 6.
39. John Bunting and Pete Hardin, “USDA’s Milk-Pricing Fails: Producers Lose Half a Billion Dollars,” *Milkweed* 332 (Mar. 2007). The article refers to events in 2006.
40. Byran W. Gould, “Consolidation and Concentration in the U.S. Dairy Industry,” *Choices*, 1999–2010, http://www.choicesmagazine.org/magazine/article .php?article=123.
41. Food and Water Watch, “Consolidation and Price Manipulation in the Dairy Industry, Mar. 2011,” http://documents.foodandwaterwatch.org/doc/Dairy Competition-web.pdf.
42. Andrew Martin, “In Dairy Industry Consolidation, Lush Paydays,” *New York Times*, Oct. 27, 2012. A different lawsuit against Dean Foods et al. that involves some of the same facts was still ongoing as of early 2014: Andrew Longstreth, “U.S. Appeals Court Rules Against Dean Foods in Antitrust Case,” Reuters, Jan. 3. 2014; *Food Lion et al. v. Dean Foods et al.*, U.S. 6th Circuit Court of Appeals, No. 12-5457.
43. Shields, “Consolidation and Concentration in the U.S. Dairy Industry,” 8.
44. Andrew Martin, “Awash in Milk and Money: Lush Paydays on the Path to Dairy Industry Mergers,” Oct. 28, 2012; USDA Economic Research Service, “Organic Market Overview,” Apr. 7, 2014, http://www.ers.usda.gov/topics /natural-resources-environment/organic-agriculture/organic-market-over view.aspx#U7x_kRaViu4.
45. Committee on the Role of Alternative Farming Methods in Modern Production Agriculture, National Research Council, *Alternative Agriculture* (Washington, D.C.: The National Academies Press, 1989).
46. USDA Economic Research Service, “Organic Market Overview”; USDA, Sustainable Agriculture Research & Education (SARE) “Community Supported Agriculture,” http://www.sare.org/Learning-Center/Bulletins/Marketing -Strategies-for-Farmers-and-Ranchers/Text-Version/Community-Supported -Agriculture.

Ten

DON'T BE CRUEL

1. Amy Hatkoff, *The Inner World of Farm Animals: Their Amazing Intellectual, Emotional, and Social Capabilities* (New York: Stewart, Tabori & Chang, 2009).
2. Montgomery, *A Cow's Life*, 199–200.
3. Marcia Endres, "What Do We Know About Cow 'Friendship'?" *Daily Star*, May 9, 2009. Reprinted on University of Minnesota Extension page, http://www.extension.umn.edu/agriculture/dairy/facilities/cow-friendship.
4. Deborah Netburn, "Study Finds Calves Raised with 'Friends' Are Smarter," *Los Angeles Times*, Mar. 2, 2014.
5. Kristin Hagen and Donald M. Broom, "Emotional Reactions to Learning in Cattle," *Applied Animal Behavior Science* 85, no. 3 (Mar. 2004): 203–13.
6. Bud Williams Schools, "Teaching Low Stress Livestock Handling Methods," http://stockmanship.com/?page_id=303.
7. D. Hanna, I. A. Sneddon, and V. E. Beattie, "The Relationship Between the Stockperson's Personality and the Productivity of Dairy Cows," *Animal* 3, no. 5 (2009): 737–43.
8. Newcastle University, "Personal Touch in Farming: Giving a Cow a Name Boosts Her Milk Production," *ScienceDaily*, Jan. 28, 2009.
9. Canadian researchers found a thousand times more pathogens in the air of confinement buildings than in outdoor air. It's not just cows that suffer: Nearly a third of people who work with confined animals have chronic respiratory problems. Bill Niman and Nicolette Hahn Niman, "For Animals, Grass Each Day Keeps Doctors Away," *The Atlantic*, May 19, 2010.
10. P. A. Oltenacu and D. M. Broom, "The Impact of Genetic Selection for Increased Milk Yield on the Welfare of Dairy Cows," *Animal Welfare* 19 S (2010): 39–49.
11. Temple Grandin and Mark J. Deesing, eds., "Genetics and Animal Welfare," Dept. of Animal Science, Colorado State University, Fort Collins, 1998, with 1999 updates, 14.
12. Temple Grandin and Catherine Johnson, *Animals Make Us Human: Creating the Best Life for Animals* (Boston: Houghton Mifflin Harcourt, 2009), 164–65.
13. Lorraine Murray, "The Big Business of Dairy Farming: Big Trouble for Cows," *Encyclopædia Britannica/Advocacy for Animals*, June 11, 2007, http://advocacy.britannica.com/blog/advocacy/2007/06/dairy-farming/.
14. EPA/Ag 101, "Lifecycle Production Phases," http://www.epa.gov/agriculture/ag101/dairyphases.html.
15. Clodagh Finn, "Big Bertha," *Independent.ie*, Sept. 29, 2012.

16. Stephen J. Hedges, "How Weak Cows Enter Food Chain: Emaciated, Calcium-Depleted Dairy Cattle Are Turned into Meat," *Chicago Tribune*, Mar. 2, 2008.
17. USDA Food Safety and Inspection Service, "Veal from Farm to Table," June 2013, http://www.fsis.usda.gov/wps/wcm/connect/c1c3ed6a-c1e5-4ad0-ba6c-d53d71d741c6/Veal_from_Farm_to_Table.pdf?MOD=AJPERES.
18. Erik Eckholm, "Farmers Lean to Truce on Animals' Close Quarters," *New York Times*, Aug. 12, 2010, http://www.nytimes.com/2010/08/12/us/12farm.html.
19. The Beef Council, "Veal Farm, Today's Veal, Veal Facts": "Individual stalls maximize the quality of individual care farmers and veterinarians can give the calves. Also, most importantly, minimizing calf-to-calf contact is the best prevention against disease," http://www.vealfarm.com/vealfacts.aspx, accessed Aug. 17, 2013.
20. American Veal Association, "Veal Farmers Move Calves to Group Pens," PORKNetwork, May 9, 2012, http://www.porknetwork.com/pork-news/Veal-farmers-move-calves-to-group-pens-150793375.html.
21. The Humane Society of the United States, "HSUS Investigation Results in Closure of Vermont Slaughter Plant," Oct. 30, 2009, http://www.humanesociety.org/news/press_releases/2009/10/vt_slaughter_plant_investigation_103009.html.
22. Greg Lardy, "Systems for Backgrounding Beef Cattle," North Dakota State University Extension Service, Aug. 2013, http://www.ag.ndsu.edu/pubs/ansci/beef/as1151.pdf.
23. Explore Beef, Modern Beef Production Fact Sheet, "The Stages of Beef Production," http://www.explorebeef.org/CMDocs/ExploreBeef/FactSheet_ModernBeefProduction.pdf.
24. Montgomery, *A Cow's Life*.
25. USDA Economic Research Service, "Cattle and Beef," May 12, 2012, http://ers.usda.gov/topics/animal-products/cattle-beef/background.aspx#.U7ySjhaViu4.
26. Johann F. Coetzee et al., "A Survey of Castration Methods and Associated Livestock Management Practices Performed by Bovine Veterinarians in the United States," *BMC Veterinary Research* 6, no. 12 (Mar. 3, 2010).
27. South Dakota State University Veterinary Extension, "Beef Cattle Procedures: Castration," http://www.sdstate.edu/vs/extension/beef-castration.cfm.
28. AMVA/Knowledge Base/Resources/Backgrounders, "Welfare Implications of Castration of Cattle," Apr. 20, 2012.
29. P. M. Faulkner and D. M. Weary, "Reducing Pain After Dehorning in Dairy Calves," *Journal of Dairy Science* 83, no. 9 (Sept. 2000): 2037–41.
30. W. K. Fulwider et al., "Survey of Dairy Management Practices on One Hun-

dred Thirteen North Central and Northeastern United States Dairies," *Journal of Dairy Science* 91, 4 (Apr. 2008): 1686–92.

31. People for the Ethical Treatment of Animals, "Casey Affleck Speaks Up for Cows Mutilated for Milk," http://www.peta.org/videos/casey-affleck-speaks-up-for-cows-mutilated-for-milk/.
32. AVMA, "Welfare Implications of Tail Docking of Cattle," May 29, 2013, https://www.avma.org/KB/Resources/LiteratureReviews/Pages/Welfare-Implications-of-Tail-Docking-of-Cattle.aspx.
33. Mark Peters, "Dairies Curtailing Cow-Tail Cutting," *Wall Street Journal*, Sept. 3, 2012.
34. Humane Society of the United States, "An HSUS Report: Welfare Issues with Tail Docking of Cows in the Dairy Industry," Oct. 2012.
35. Gail A. Eisnitz, *Slaughterhouse: The Shocking Story of Greed, Neglect, and Inhumane Treatment Inside the U.S. Meat Industry* (New York: Prometheus Books, 1997); Alex Hershaft, Review of *Slaughterhouse*, by Gail A. Eisnitz, http://www.britishmeat.com/slaught.html; Donald McNeil, "Videos Cited in Calling Kosher Slaughterhouse Inhumane," *New York Times*, Dec. 1, 2004; PETA, "Agriprocessors Fined $10 Million: Sholom Rubashkin Arrested," Oct. 30, 2008, http://www.peta.org/blog/agriprocessors-fined-10-million-sholom-rubashkin-arrested/.
36. Helena Bottemiller, "Landmark Settlement Reached in Westland-Hallmark Meat Case," *Food Safety News*, Nov. 18, 2012, http://www.foodsafetynews.com/2012/11/landmark-settlement-reached-in-westlandhallmark-meat-case/#.U43_n_ldXa4.
37. Mark Bittman, "Who Protects the Animals?" *New York Times* op-ed, Apr. 27, 2011.
38. Cheryl Hanna, "Animal Protection Organization Mercy for Animals Reveals Idaho Dairy Cow Abuse," Examiner.com, Oct. 11, 2012; NBC-News.com, "Torment of Dairy Cows in Undercover Video Leads to Cruelty Charges," Oct. 10, 2012.
39. Baylen Linnekin, "How Ag Gag Laws Suppress Free Speech and the Marketplace of Ideas," Reason.com, Sept. 1, 2012.
40. Ken Broder, "Undercover Video Exposes Illegal Abuse of Animals Used in School Lunch Program . . . Where Were USDA Inspectors?" AllGov.com, Aug. 23, 2012.
41. M. L. Johnson, "DiGiorno, Supplier Drop Dairy Farm over Abuse," Associated Press/*USA Today*, Dec. 10, 2013.
42. Temple Grandin, "Return-to-Sensibility Problems After Penetrating Captive Bolt Stunning of Cattle in Commercial Beef Slaughter Plants," *Journal of the American Veterinary Medical Association* 221. no. 9 (Nov. 1, 2002): 1258–61.

43. Jennifer Mascia, "Animals, Cruelty and Videotape," *New York Times*, Apr. 27, 2011.
44. Dan Flynn, "Five States Now Have 'Ag-Gag' Laws on the Books," *Food Safety News*, Mar. 26, 2012, http://www.foodsafetynews.com/2012/03/five-states-now-have-ag-gag-laws-on-the-books/#.U4zBlvldXa4.
45. Nicolette Hahn Niman, "Support Your Local Slaughterhouse," *New York Times* op-ed, Mar. 2, 2014. Rancho Feeding Corporation has since been acquired by Marin Sun Farms Inc., which focuses on sustainable meat.
46. Department of Labor, Bureau of Labor Statistics, "51-3023 Slaughterers and Meat Packers," May 2012, http://www.bls.gov/oes/current/oes513023.htm.
47. Eric Schlosser, *Fast Food Nation* (New York: Perennial, 2002), 173.
48. Eric Schlosser, "How to Make the Country's Most Dangerous Job Safer," *The Atlantic*, Jan. 2002.
49. The Humane Society of the United States, "Rampant Animal Cruelty at California Slaughter Plant: Undercover Investigation Finds Abuses at Major Beef Supplier to America's School Lunch Program," Jan. 30, 2012, http://www.humanesociety.org/news/news/2008/01/undercover_investigation_013008.html.
50. Margaret Shakespeare, "At Home on the Range," *Stanford*, July/Aug. 2013, 49–55.
51. For example, see Puget Sound Meat Producers Cooperative, http://www.pugetsoundmeat.com/. An explanatory video of the Ranch Foods Direct unit, which is Animal Welfare Approved, may be found at http://www.mobilemeatprocessing.com/.
52. USDA, "Bull Management Practices on U.S. Beef Cow-Calf Operations," Feb. 2009, http://www.aphis.usda.gov/animal_health/nahms/beefcowcalf/downloads/beef0708/Beef0708_is_BullMgmt.pdf.
53. Doug Erickson, "Noted University of Wisconsin–Madison Researcher Henry Lardy Dies at 92," *Wisconsin State Journal*, Aug. 6, 2010; Nobelprize.org, "Paul D. Boyer—Biographical," http://www.nobelprize.org/nobel_prizes/chemistry/laureates/1997/boyer-bio.html.
54. John H. Tobe, *Milk: Friend or Fiend?* (St. Catharines, Ontario: Modern Publications, 1967), 11.
55. Martin LaMonica, "'Smart Cows' Can Text Owners When Mooood Strikes," CNET News, Feb. 22, 2012.
56. Roger L. Davis, "Embryo Transfer in Beef Cattle," Davis-Rairdan International (Alberta, Canada, ca. 2004); Ross Wilson, "Embryo Transfer in Cattle," Cruachan Highland Cattle Society, 1992, http://www.davis-rairdan.com/embryo-transfer.htm.
57. Although a clone has DNA identical to that of its parents, it won't be an identical animal. For example, cloned Holsteins don't have exactly the same

pattern of spots or the same ear shape as their parents. This is because non-genetic influences can act upon DNA to turn genes on and off.

58. Jenifer Horton, "5 Most Cloned Animals," Discovery Channel (undated—ca. 2008), http://www.discovery.com/tv-shows/curiosity/topics/5-cloned-animals.htm.
59. FDA, "Myths About Cloning," Oct. 28, 2009, updated Oct. 28, 2009, http://www.fda.gov/AnimalVeterinary/SafetyHealth/AnimalCloning/ucm055512.htm.
60. Pew Charitable Trusts Health Initiatives, "Americans' Knowledge of Genetically Modified Foods Remains Low; Majority Are Skeptical About Animal Cloning," press release, Nov. 15, 2005.
61. Consumers Union, "CA Bill: Require Labeling on Cloned Food," Jan. 24, 2008, http://consumersunion.org/news/ca-bill-require-labeling-on-cloned-food/.
62. Martha Rosenberg, "Cloned Meat May Already Have Invaded Our Food Supply, Posing Alarming Health Risks," *AlterNet*, Aug. 19, 2010.
63. Karen Kaplan, "FDA Standing in the Way of Montana Rancher's Leap into Cloning Revolution," *Los Angeles Times*, Feb. 15, 2005. Kaplan reports cloned calves being raised by youngsters. The calves would go into the human food chain after prancing around the show ring.
64. Center for Food Safety, "Groups Tell FDA, Keep Food from Cloned Animals Off Our Dinner Plates," Oct. 2006, http://www.centerforfoodsafety.org/press-releases/888/groups-tell-fda-keep-food-from-cloned-animals-off-our-dinner-plates.
65. P. Chavatte-Palmer et al., "Health Status of Cloned Cattle at Different Ages," *Cloning Stem Cells* 6, no. 2 (2004): 94–100; G. Vajta and M. Gjerris, "Science and Technology of Farm Animal Cloning: State of the Art," *Animal Reproduction Science* 92, nos. 3–4 (May 2006): 211–30.
66. D. N. Wells, "Animal Cloning: Problems and Prospects," *Revue Scientifique et Technique* (International Office of Epizootics) 24, no. 1 (2005): 251–64.
67. FDA, "Myths About Cloning."
68. D. N. Wells et al., "The Health of Somatic Cell Cloned Cattle and Their Offspring," *Cloning Stem Cells* 6, no. 2 (2004): 101–10.
69. In the late 1990s, Tom Elliott sold the N Bar. Sinclair Cattle Co. Inc. bought his two hundred best cows and semen/embryo inventory.
70. Jenny Kirk, "Cow Comfort: Wabasso Brothers Use Cutting-Edge Technology at Their Family Dairy Farm That Has Improved the Health of Their Cows and Gotten Worldwide Notice," *Marshall Independent*, Apr. 21, 2012.
71. J. A. Jacobs and J. M. Siegford, "Invited Review: The Impact of Automatic Milking Systems on Dairy Cow Management, Behavior, Health, and Welfare," *Journal of Dairy Science* 95, no. 5 (May 2012): 2227–47.

72. Jeremy Hainsworth, "Her Secret to Tasty Beef? Buckets of Wine," *Seattle Times*, Sept. 12, 2010.
73. Nick Carbone, "Loaded Livestock: French Farmers Serve Cows 2 Bottles of Wine per Day," *Time* NewsFeed, July 14, 2012.
74. Alex Paul, "Van Loon Dairy Cows Enjoying Waterbed Pampering," *Albany Democrat-Herald*, July 9, 2012.
75. Abbey Gibb, "Oregon Farms Try Out Waterbeds for Cows," KGW.com, Portland, OR, July 16, 2012 (updated Dec. 18, 2012).
76. Holly Richmond, "These Cows Frolicking in a Field are Guaranteed to Make You Smile," *Grist*, Aug. 22, 2013.
77. Fox News.com, "Drought Crushes Hawaii Ranchers' Efforts to Meet Demand for Locally Raised, Grass-Fed Beef," Oct. 4, 2012.
78. *Visitmolokai.com Visitor Center*, http://visitmolokai.com/faq.php#deer1, citing the *Hawaiian Gazette* of Dec. 17, 1867.
79. *Honolulu Star Advertiser*, "New Law Prohibits Having or Releasing Feral Deer in Hawaii," June 22, 2012; *Honolulu Star Advertiser*, "Deal Humanely with Axis Deer," editorial, Feb. 8, 2011.
80. Victoria W. Keener et al., eds., *Climate Change and Pacific Islands: Indicators and Impacts: Report for the 2012 Pacific Islands Regional Climate Assessment (PIRCA)* (Washington, D.C.: Island Press, 2012), Executive Summary.

Afterword
THE TAIL END OF AN ERA

1. Sara Schoenborn, "Per Capita Meat Consumption Predicted Lower for 2013; Prices Moderately Higher," *Agri-View*, Feb. 21, 2013, http://www.agriview.com/news/livestock/per-capita-meat-consumption-predicted-lower-for-prices-moderately-higher/article_7338d748-7c38-11e2-bc63-0019bb2963f4.html.
2. "Dinner Is Served," http://www.monticello.org/site/jefferson/dinner-served.
3. Local Harvest is a group that connects farmers with people looking for high-quality, locally produced food. The organization's website has national lists of small farms and farmers' markets, and information for both farmers and consumers on CSAs, http://www.localharvest.org/about.jsp.

Index

Page numbers in *italics* refer to illustrations.